An un-birthday,

un-anniversary,

un-mother's or
father's day

present

-- love, Larry
+
Marg.

August, 1978

The New Oxford Book of Light Verse

The New Oxford Book of Light Verse

Chosen by
Kingsley Amis

OXFORD UNIVERSITY PRESS
Oxford London New York

Oxford University Press, Walton Street, Oxford OX2 6DP

OXFORD LONDON GLASGOW NEW YORK
TORONTO MELBOURNE WELLINGTON CAPE TOWN
IBADAN NAIROBI DAR ES SALAAM
KUALA LUMPUR SINGAPORE JAKARTA HONG KONG TOKYO
DELHI BOMBAY CALCUTTA MADRAS KARACHI

British Library Cataloguing in Publication Data

The new Oxford book of light verse.
1. Humorous poetry, English
I. Amis, Kingsley
821′.07 PR1195.H8

ISBN 0–19–211862–5

PRINTED IN GREAT BRITAIN BY
RICHARD CLAY (THE CHAUCER PRESS) LTD
BUNGAY, SUFFOLK

INTRODUCTION

I

I must now attempt what I hardly had to think of at all, except as a rather daunting prospect, while I was putting this anthology together, and try to define light verse, or at least propose some general remarks about it that shall bear on my selection.

My illustrious predecessor was not much help to me here. His choice of poems and mine had differed widely, so much so that, for instance, our respective extracts from *Don Juan* turned out not to have one stanza in common. And a book of light verse that includes 'Danny Deever', one of the most harrowing poems in the language (nothing else by Kipling, either), must be founded on principles quite unlike mine.

These—Auden's principles—are asserted in his Introduction to have to do with the kind of poet who unselfconsciously shares the common life and language of ordinary men and writes of the one in the other, in something close to the speaking voice: what he then produces will be light. This justifies the large selection of folk-songs, ballads, blues lyrics (for all their recurrent themes of betrayal and murder) and other pieces of straightforwardness, though that 'unselfconsciously' ought to have ruled out 'Danny Deever' and much else. The further Auden moves forward in time, the wider the gap between statement and practice (or reality). 'Light verse tends to be conventional' and conservative. Rochester? Byron?—only if we assume that nobody born in 'Society' can ever reject it. Auden himself?

Of all people, Dryden and Pope—I pause over them for a moment to help to justify two of my own omissions—are described as able 'to use the speaking voice, and to use as their properties the images of their everyday, i.e. social, life'. But, if weight is a useful opposite of lightness, no English poet is heavier than the author of, say, *Absalom and Achitophel*, further from conversational tones; more of it later. And, although it would be widely agreed that light verse does use everyday properties, Pope could not do so for long without elevating them out of their nature, putting China's earth into an unmentioned teapot and transforming combs back into their primordial tortoise and elephant. The verse, too, far from constantly suggesting that both it and the heroic couplet in general have something ridiculous about them, as the verse of light verse would do (compare the *ottava rima* of *Don Juan*), is very well pleased with itself.

Auden's political preoccupations in those days (1938) led him in conclusion, with what accuracy we can now judge, to foresee a planned egalitarian state of society in which light verse would flourish as never (well, hardly ever) before. But I have already gone quite far enough into this argument. I might not have gone so far, had it not been for a view expressed by Mr. Peter Porter, an authority on the subject and one who has given me generous help in compiling the present volume. Mr. Porter called Auden's a 'revolutionary anthology' and said that it 'chang[ed] the sensibility of a generation'. I wonder what generation he had in mind. My own, as far as I can tell, stayed largely untouched by it, even niggling a little at the choice of jazz and American folk lyrics, though we, or I, have found it an indispensable source of much curious and out-of-the-way material, 'light' or not. The principle of checks and balances, and I, will be satisfied if another generation altogether sees in mine a reactionary anthology.

II

As will be noted, Auden said a good word in 'Letter to Lord Byron' for A. A. Milne as a writer in the genre, though he was no more able than I to find a specimen worth anthologizing. Certainly Milne gave by far the best account of light verse I have come across in his *Year In, Year Out* (1952). He did so in the course of an essay on C. S. Calverley, whom he called the 'supreme master' of light verse. It puts so many necessary points so exactly and elegantly that I must quote from it at length.

Light Verse obeys Coleridge's definition of poetry, the best words in the best order; it demands Carlyle's definition of genius, transcendent capacity for taking pains; and it is the supreme exhibition of somebody's definition of art, the concealment of art. In the result it observes the most exact laws of rhythm and metre as if by a happy accident, and in a sort of nonchalant spirit of mockery at the real poets who do it on purpose. But to describe it so leaves something unsaid; one must also say what it is not. Light Verse, then, is not the relaxation of a major poet in the intervals of writing an epic; it is not the kindly contribution of a minor poet to a little girl's album; it is not Cowper amusing (and how easily) Lady Austin, nor Southey splashing about, to his own great content, in the waters of Lodore. It is a precise art which has only been taken seriously, and thus qualified as an art, in the nineteenth and twentieth centuries. It needs neither genealogical backing nor distinguished patronage to make it respectable.

From time to time anthologies of light verse are produced. The trouble with most of the anthologists is that, even if they have an understanding of their subject, secretly they are still a little ashamed of it. They try to give it the blessings of legitimacy by tracing its ancestry back to some dull fourteenth-century poem beginning '*Lhude sing cuccu*' or '*Merry swithe it is in halle*'. If any intervening major poet (other than Byron) has unbent for a moment, the distressing result is dragged in, to provide the sanction of respectability that the Vicar brings to the raffle at the Choirboys' Outing. 'Poets at Play' said one anthologist, and we see at once there is no real harm in it. One mustn't take them seriously, it's only their fun.

Now if anybody wishes to see what happens when a poet is at play, he has only to read that most deplorable piece of doggerel, *John Gilpin*. Light Verse is not the output of poets at play, but of light-verse writers (who would not thank you today for calling them poets) at the hardest and most severely technical work known to authorship. It is not bastard poetry on a frivolous theme. It is true humour expressing itself in perfectly controlled rhyme and rhythm . . . .

Cowper wrote:

> Quoth Mrs. Gilpin 'That's well said;
>   And for that wine is dear,
> We will be furnished with our own
>   Which is both bright and clear.

Of course Mrs. Gilpin quoth nothing of the sort. When recommending a wine, one advances something more in its favour than that it is 'both bright and clear'; nor was it likely that Gilpin knew less about the wine in his cellar than his wife, and needed the information. What she really said (in prose) was: 'A very good idea; and since all these inns, and I don't suppose the Bell is any exception, overcharge ridiculously, we'd better take our own wine with us.' Cowper had to translate this into verse. In the result the words are unnatural, the scansion is strained, and the first and third lines do not rhyme, as in light verse they should. In fact, the speech as translated has nothing to recommend it, save that it can just qualify as verse by reason of having the right number of feet and a rhyme for the second and fourth lines. But it is not light verse. It is very heavy verse. We are told that Cowper 'jotted it down during a sleepless night', 63 verses of it. It bears all the signs of having been jotted down.

I have now amused myself by making my own translation of Mrs. Gilpin's speech. Here it is. It has faults of its own, no doubt: an excess of frivolity, perhaps: but it obeys the rules. It is 'light verse' in its own right; not merely light because the subject is light, or because I didn't work at it seriously.

> Said Mrs. Gilpin: 'Very well.
>   But wait a moment! What'll
> They charge for claret at the Bell?
>   We'd better take a bottle.'

Naturally, I have some reservations. I share Milne's admiration for Calverley, but cannot feel that he has no equals. And of course I would contest the view that, Byron apart, major or at least 'real' poets have unbent to no advantage except that of legitimizing light verse, and would adduce a couple of dozen of those represented in what follows.

None the less, Milne's rewriting of Cowper's stanza summarizes a great deal.

III

'All good poetry has something light about it,' said Mr. Porter in a broadcast, though a moment later he was obliged to except Milton. This is a recent version of the demand made by somebody very different from Mr. Porter, namely D. G. Rossetti, that poetry should be as 'amusing' as any other product of the mind. Here we have useful guidance for writers of high verse, but not much is being said about light verse. The two are different arts with a strong one-sided relationship. High verse could exist without light verse, however impoverished life would be if that were the state of affairs. But light verse is unimaginable in the absence of high verse. We are told that all poetry refers to all the other poetry extant at the time of composition. With high verse this reference will usually be distant, often imperceptibly tenuous; with light verse it is intimate and essential. To this degree it is altogether literary, artificial and impure.

Light verse makes more stringent demands on the writer's technique. A fault of scansion or rhyme, an awkwardness or obscurity that would damage only the immediate context of a piece of high verse endangers the whole structure of a light-verse poem. The expectations of the audience are different in the two cases, corresponding to the difference in the kind of performance offered. A concert pianist is allowed a wrong note here and there; a juggler is not allowed to drop a plate.

The obvious opposite of 'high' is 'low', and light verse is low in both the Augustan and the Victorian senses; it is—again doubly—offensive to decorum. It can only be dignified as a preparation for bathos. Unless engaged in parody, it prefers forms incompatible with decent seriousness: jogging rhythms, elaborate rhymes, stanzas that erect trip-wires for the unwary reader. It deals with low matters, with subjects, scenes and concerns that are either poetically or morally unsuitable for high consideration. It uses low terms, whether rustic, technical, colloquial, facetiously anachronistic, or vulgar, ill-bred, obscene. Its chief weapon is impropriety.

All this sounds solemn enough; it is time for me to remember that light verse is *light*, light in the sense of cheerful, airy, light as in light-footed and light-hearted, 'requiring little mental effort; amusing, entertaining' (OED). The writer of that dictionary article was not concerned to say that it is the audience, not the performer, from whom that little mental effort is called for, nor that, in the case of light verse at least, mental relaxation in the one will leave undiscovered a large part of what the other provides. Just looking through Praed, which his mastery makes so easy and so tempting, misses almost everything he was about. And perhaps this is the place—I have to find a place—to say that all light art is likely to deliver, now and then, a jolt to the gentler emotions, the more telling for its unexpectedness. Light music is always doing this: how many lunch-time listeners to broadcasts of Johann Strauss have not found themselves, at certain moments, suddenly unable to go on chewing? With light verse, this effect is obviously most likely to come near the end of the piece. At such times, I should argue, the verse ceases to be light, the poet's tone of voice begins to belong to that of high verse, and—if one imagines him reciting in public—his manner would correspondingly change. This happens, for instance, in the final stanza of Chesterton's 'Rolling English Road'. It also happens with the third of Leigh Hunt's fish sonnets, here included on the editorial principle that a unity that marries light and high verse so closely and to such effect must not be broken. That third sonnet, by the way, is beautiful, which light verse, unlike light music, can never be. Genial, memorable, enlivening and funny is the loftiest description it can aspire to.

That is something, and it leads me to return to the point from which I have a little digressed. It has been urged more than once that the inability of light verse to be beautiful is one result of the weakening that English poetry as a whole underwent some time after the middle of the seventeenth century. Before then, the story runs, there was a single style in which the poet could move from grave to gay and back again with perfect ease and congruity; for Auden, it was a light style. Aldous Huxley had put the point five years earlier in his *Texts and Pretexts*, a historically interesting anthology-with-commentary. Although Huxley had not much feeling for poetry and no sense of humour—a frequent combination, especially among literary critics—his account is representative. ('Comic' is not everywhere interchangeable with 'light' when verse is under discussion, but it is here.)

Compared with the best Elizabethan specimens, the comic poetry of later times seems, even when actually wittier and more amusing, rather poor stuff. Poor in not being beautiful. A certain natural and easy eloquence distinguished the comic verse of the Elizabethans, just as it distinguished their serious verse. Their fun is in the grand manner. Whereas ours is, and for the last two centuries has been, in the flippant manner—flippantly low, or else flippantly too high, mock-heroic. We make a radical distinction between the comic and the serious style. Which is a profound mistake. The best comic works have been grand and beautiful . . . [He actually gives Rabelais as an example.] The Elizabethans used the same style (in their case a rich and musically flowing one) for both kinds of poetry.

Unfortunately for his argument, Huxley supplies three examples of 'comic verses that are also beautiful'. The second of these is a well-known anthology piece usually given as anonymous, though Auden, whose text I follow, ascribes it to Thomas Weelkes (1575?–1625). The middle stanza of three runs:

> Tie hie! tie hie! O sweet delight!
> He tickles this age that can
> Call Tullia's ape a marmasyte
> And Leda's goose a swan.
> Fara diddle dyno,
> This is idle fyno.

Idle something, anyway. No doubt it was an age easily tickled. To leave Tullia's ape out of it, the bird associated with Leda is a swan, not a goose. Weelkes presumably meant something like 'Call Audrey's goose Leda's swan', but that was too hard for him. 'You could not ask for a prettier cadence than' the three lines that mean something, says Huxley. Oh yes I could: I should ask the poet to move or get rid of that 'can'. With the heavy emphasis it picks up from its position at the end of the line, it carries an inappropriate suggestion of 'just about manages to, turns out to have been able to when everyone was betting he wouldn't be'. High verse might possibly carry this off as indicating breathlessness or stumbling sincerity; the writer of light verse must pay rigorous attention to, because he stands to be far more gravely damaged by, such small matters. The falsity of the rhyme can/swan is likely but not certain.

The wordless noises—'tie hie!' and the rest; the other stanzas have 'ha ha!' and 'so so!'—are doing the same job as the music-hall comedian's red nose and check trousers: making quite sure the audience knows that the performer is trying to be funny. 'Idle' is a hedging of bets: if you don't think it's funny after all, it doesn't matter because the poet was trifling from the start. He is at play; one mustn't take him seriously. Perhaps I should have said earlier that Huxley despised Gilbert.

Light verse need not be funny, but what no verse can afford to be is unfunny, a proposition that disposes with amazing speed of most of the earlier material commonly proffered as light or comic by anthologists. Such parts of 'The Miller's Tale' or 'The Wife of Bath's Prologue' as do not collapse under this objection are either high verse or nothing in particular. I will say no more about Dunbar than that he wrote in the period after Chaucer. I think Skelton is unfunny; I am not sure, because he is such a difficult poet; either way he gets his discharge, for light verse must not, cannot be difficult. The remainder really comes down to the anonymous and the doubtfully attributed.

Anon. is not my favourite poet. If he were good or very good, I feel he would have a full name and a pair of dates. (Of course the case is different when he might have chosen anonymity to avoid political or puritanical disapprobation.) He is incapable of sustained flights, preferring the song, of the kind that consists largely of refrains and repetitions, and the epigram, that perilous form. For obvious reasons, his work tends to lack the careful finish required by light verse. And he has his own remarkably well-sustained line in unfunniness, nowhere better exemplified than in those over-familiar drinking-songs of the sixteenth century—'Bring us in Good Ale', 'Back and Side Go Bare'. Some mysterious revenge of nature has seen to it that no poem in praise of drink or tobacco (or snuff, if any) can succeed.

In the next century there came the great dissociation. The date, if a date can be put on such a thing, is 1660, when Charles II and his Court returned from France. The King had spent some years in Paris, including those of his later education, and it was only natural that his taste in poetry, among other things, had become French—or, better, Frenchified. Monarch and Court in those times determined such matters far more completely than any single force does today; English poets, then, were encouraged to look to French models and mentors. The most influential of these was probably Nicolas Boileau, the 'legislator of Parnassus', whose prescriptions were for elegance, regularity and correctness in the (vague) spirit of classical Latin verse. Dryden, appointed poet laureate in 1668, set out to supply these totally alien qualities, and the Augustan Age had begun.

This old-fangled story, which I have compressed as much as I dare, has a special significance for us. Under Dryden English poetry became elevated, stilted in the original sense of 'on stilts', high verse with a vengeance. Its social counterpart, a new thing for it to have, was a supposed refinement and polish of manner: in his critical works, Dryden was fond of commending a piece of writing for its 'courtly' style or its reflection of the author's 'conversation [i.e. social intimacy, not just chatting] with gentlemen'. Dr. Leavis summarized the process neatly in the days when he was still talking sense: 'With Dryden begins the period of English literature when form is associated with Good Form.' It was time for another kind of poetry, subversive, disrespectful and ungentlemanly, to strike back.

The foregoing is likewise not altogether new. Nor is what follows, though its application to the matter in hand may sometimes be overlooked. If it took the Restoration and its cultural consequences to invite this upsurge, how did it come about that light verse was in such an advanced state of preparedness? The first part of *Hudibras*, for instance, was published in 1663, when Dryden had scarcely started. The answer must be that the tradition of English poetry, from Chaucer to the Elizabethans, had suffered two earlier attacks. One had come from the Metaphysicals, first and foremost Donne, whose poems had begun to be published in 1633, though they had circulated widely in manuscript copies much earlier. I am not reprehending him or his successors when I notice that the Metaphysical style strained the language, forced it into behaviour much more unnatural than anything Shakespeare imposed upon it, took it far away from Auden's 'speaking voice'. The second attack had come from Milton, the only English poet no one could call light in even the remotest sense. *Paradise Lost* was not to be published till 1667, but the 'minor' poems, including especially 'Comus' and 'Lycidas', had been in print for thirty years by then. It is enough to say of them here that, to mention nothing more than their vocabulary, syntax, word-order and verse-movement, they are unlike anything else in the language up to that point.

At this stage I had better make it clear that my talk of striking back, of a state of preparedness for doing so and of attacks on the tradition should not be taken too literally. It was not that Dryden, Milton and Donne (or Cowley, one of the last Metaphysicals and a contemporary of Milton's) set about a purposeful attack, together or separately, on something they could have defined as an existing tradition, nor that, say, Butler and Rochester saw what was happening and elected to counter-attack. It cannot have been as self-conscious as that. But surely the innovators felt themselves to be innovators, those who reacted had sensed, vaguely perhaps but not therefore less strongly, that there were new or recent poetic styles which were asking to be made to look odd or obscure or pompous or exclusive (the last is the best I can do in the way of a word meaning 'such as to exclude a great deal of subject-matter and some moods and tones that had always been admissible before'). If this is still too rationalized for you, try saying that English poetry had now reached a stage at which there was something there to be light in comparison and contrast with.

A sufficient variety of forms lay ready to hand, from the loosest ballad or song metre to the strictest simulacrum of high verse in heroic couplet or decasyllabic quatrain. With this there went an equally wide range of manner or method, from satire, which for the most part is aggressively sensible, to what is called nonsense: I will discuss in a moment some of the main categories. It is worth stressing the fact of this variety and range. Light verse is not one thing but many, so much so that I should hate to have to frame a single generalization which would comprehend them all: every law I have laid down admits of the odd exception. Rough rules and limitations, some of which may seem personal or even idiosyncratic, are a different matter. They will appear as I proceed.

IV

'To raise a good-natured smile was the major part of this work written,' said Charles Dibdin in the preface to his *Comic Tales and Lyrical Fancies* (1825). We are not far from a handy short definition of light verse here. The smile need not come as a response to the comic; it may acknowledge one of several sorts of outrageousness, a parodist's exact catching of his original's tone of voice, a stroke of metrical ingenuity, etc. The kind of smile is most important. Going to elaborate lengths in order to raise an ill-natured smile seems an unworthy enterprise, one moreover that conflicts with most associations of the word 'light'. It is what a great deal of satire is up to, including some of the most accomplished technically. I have omitted pieces written in this spirit.

Hence, or partly hence, the absence of Pope, who may have been possessed by the strong antipathy of good to bad, but whose antipathy to those who had irked him was quite as strong and more immediately noticeable. His victims were so many that they included hundreds of people now—no doubt deservedly—forgotten. It is normally an editor's task to provide notes identifying these and perhaps indicating what they had done to arouse Pope's displeasure. The need for such informative distractions damages the work concerned and their presence in a book of light verse is an intrusion. At any rate, there are none here, and those poems that need them are missing too.[1]

Dryden is a rather different case. It would have taken a greater poet than him to have written at length in a style inimical to another which he had himself perfected, and *Absalom and Achitophel*—the test case, already mentioned, a great poem and one of the cleverest in the language—is nearly all high verse with what amount to accidental light-verse connections. The enterprise was a dubious one from the start: a satire on the monarch's enemies by a poet laureate, and one intended to influence events, too. ('Achitophel' was awaiting trial for high treason at the time. It is no thanks to Dryden that he got out and was able to flee to Holland.) In immediate reference to Pope, but in

[1] Not only Pope's but, among others', Charles Churchill's and Burns's, the one a good, the other a great poet. I would have broken my no-notes rule for that other if there had not been, as in Pope's case, prior grounds for excluding him. Part of Burns's very greatness lies in his distinctive tone or manner, which, while not high and as far as possible from being heavy, is still not light, but carries instead what I will call a genial gravity.

the general context of the Augustan Age, Mr. Robert Graves once wrote, 'Professional standards in English, as opposed to Continental, poetry, have for centuries insisted that satire should not reflect private rancour [not only Pope but also Dryden, who was following up a literary vendetta in part of the poem] or hope of personal gain [very much Dryden, with his financial dependence on the King].'

Given these circumstances, the result was a triumph: witty, epigrammatic, vigorous, compressed, vivid. But the tone is painfully pious. The poet is righteously indignant with Achitophel and his party, whose characters and activities he censures more in sorrow than in anger; he is nothing if not fair-minded—'The Statesman we [no egotism about *him*] abhor, but praise the Judge'; whatever the provocation, he will remain cool, detached, *lofty*. Now and again, out of an amiable weakness, he lets through a genuinely funny, spiteful remark:

> Beggared by fools, whom still he found too late,
> He had his jest, and they had his estate,

but the poet can obviously not afford to continue in that strain. In this respect, that both could write light verse but chose to do so for no more than a few lines at a time, Dryden is very like Pope. If the point needs reinforcing, let me quote from Defoe's treatment of Dryden's patron:

> The royal refugee our breed restores,
> With foreign courtiers, and with foreign whores:
> And carefully repeoples us again,
> Throughout his lazy, long, lascivious reign. . . .

and so on. *That* is the utterance of light verse.

Rather as the law finds nothing libellous in whatever might be written against an indeterminate group as distinct from an individual, so the smile raised by a satire on the Dutch, Scotch or Irish is not ill-natured. The same is true when the target is a profession, a sect, a political party or a social class or stratum. With a death-date in the eighteenth century, satire was the first light-verse form to disappear, resigning something of its function to *vers de société*. In modern similitudes like 'Caledonia' and 'Hibernia' the element of pastiche is at least as strong as that of satire.

V

There is only one thing that could be learnt by attempting to parody a writer whose distinction makes him worth close study; that is, how inaccessible to any but the most superficial, and falsifying, imitation the truly characteristic effects of such writers are. . . . What does the parody of Wordsworth in *Rejected Addresses* tell us about the great Wordsworth? Nothing . . . The cult of parody, in fact, belongs to that literary culture (a predominant one, to judge by our intellectual weeklies—it is a branch of 'social civilisation') which, in its obtuse and smug complacency, is always the worst enemy of creative genius and vital originality. It goes with the absurd and significant cult of Max Beerbohm. . . . People who are really interested in creative originality regard the parodist's game with distaste and contempt.

[Parody] is a trick not far removed from punning; yet, when well executed, it gives pleasure, I think, to any one not born a prig.

The proponent of the first of these views of the case is, perhaps unexpectedly, Dr. Leavis; the second is to be found in the Introduction to *Parodies and Imitations Old and New*, ed. J. A. S. Adam and B. C. White, 1912. On the whole, I am inclined to the second view. It is quite true that *Rejected Addresses*, like the verse in *The Anti-Jacobin* and all other purposeful or systematic parody, tells us nothing about anything, because it is no good. But no parody of any kind can be expected to tell us anything about the great Wordsworth or the great anyone else or even the fair-to-middling anyone else; that is not the point of the parodist's game, to use the phrase in a different sense. The 'cult' of Max Beerbohm, spoken of by Dr. Leavis with a horror more appropriate to a cult of Moloch, had surely been wound up long before 1962, the date of my quotation, if indeed it had ever—but let that pass. Beerbohm, a much overrated parodist in my view, did Henry James and Conrad among others, and Dr. Leavis goes on to say that they 'suffered' in consequence, evidently by losing or failing to acquire some measure of critical esteem. Really? Among poets, Swinburne must have been parodied more widely, successfully and irreverently than any, but if he suffered by it he seems not to have known, or perhaps just not to have cared, judging from his having written and published a not very effective, but not at all respectful, parody of himself.

People who are really interested in creative originality (or vital genius) will be the poorer if they disregard parody, which even at its worst and most unfriendly leaves intact the author of the original. Like so much light verse, it is an exercise in virtuosity; the writer plays on a set of tensions, between closeness to the model and distance from it, the plausible and the entertaining, what makes a fair parody and what makes a good poem in its own right. When well executed, it gives pleasure not simply to non-prigs but again and again to those otherwise unacquainted with the original and even to those altogether unaware that what they are reading is a parody, though they will know that it is light verse. You soon get your bearings and settle down. What proportion of children and others who enjoy 'You Are Old, Father William' have ever heard of Southey, or would be much better off after a look at 'The Old Man's Comforts and How He Gained Them'?

The last is no doubt an extreme case, but there are plenty of intermediate ones. Chesterton's parody of Yeats is a nice unsensational example. Almost nothing would be lost if the name of some forgotten Irish trifler were substituted at the head, though one would not feel quite safe in leaving it unsignposted: it would not be such a good poem if nobody could ever have been enough of a bloody fool to mean it seriously. A more overtly hostile piece, Owen Seaman's 'Birthday Ode', was included in an anthology of fifty years ago only with reluctance, because Alfred Austin was fading from view. Now he is gone altogether, but the poem still stands; the reader needs only to know what is proclaimed, that Austin was a bloody fool of a rather different sort. The editor of that anthology, J. C. Squire, left out Lewis Carroll's 'Hiawatha' on the grounds that 'cameras are now commonplaces'. That kind of thing never matters in itself: everything depends on how it's done—saying which is a warning to move along.

VI

Nonsense verse naturally comes next, because a great deal of what passes for nonsense is or was generic parody, that which hits not at an individual or even a group but at a kind of poetry, however dimly visualized. Thus 'Oh that my lungs' looks as if it had been touched off in the first place by impatience with the sense-confounding styles of the early seventeenth century, Fulke Greville's no less than Donne's. It is easy to forget, though just as easy to remember, that 'Jabberwocky', often taken as setting the standard for purity of nonsense, doubles as a good-natured glance at the kind of ballad that, affectedly or not, uses archaic or dialect words unfamiliar to the ordinary reader. (Nonsense, by the way, ought to yield no sense, but the narrative here can be followed easily enough even without Carroll's glossary. How could an intelligent little girl like Alice feel like confessing 'that she couldn't make it out *at all*?' [my italics] She immediately shows that she can make out the main point of it before interrupting herself and rushing out into the garden.)

Pure nonsense, or what comes closest to it while using almost nothing but ordinary English forms of language, is to be found in certain nursery rhymes: 'The Sow Came In', 'Rub-a-dub-dub' (though an earlier version shows the butcher and his two colleagues in a less nonsensical, more unseemly position), and others I have not included, like 'Old Mother Hubbard'. The 'cult' of this fribbling tale—

the toy book of it sold more than ten thousand copies within a few months of its publication in 1805—surprises and, though almost imperceptibly, depresses me. The following is not an unfair sample:

> She went to the alehouse
> To get him some beer;
> But when she came back
> The dog sat in a chair.

But? Sitting in a chair is a popular and very sensible course of action when beer is produced, from an alehouse or anywhere else. What would Mother Hubbard or the poet have had the dog do? What of it that he sat in a chair instead? I dare say he did sit in a chair. A little of 'Old Mother Hubbard', and there are fourteen stanzas of it in the received version, goes a long way with me. Like many nursery rhymes, it is whimsical, or nonsense in the derogatory meaning of the word.

Nearly all Lear's Nonsense Songs are whimsical to the point of discomfort. Louis MacNeice wrote in 1953 that they were a fad of the 1920s, which is easy to believe of a decade that produced the Sitwell–Walton *Façade*. The Songs are the utterances of that funny-bachelor-uncle figure so appealingly described, though to our good fortune not allowed to recite, in 'How Pleasant to Know . . .'—it would have been a little pleasanter still if he had had the strength of mind to forgo the pancakes and lotion, and even more those chocolate shrimps. (Manufacturers do make chocolates moulded in the shapes of various creatures, but they are not normally available at any kind of mill.) I could wish that the lands where the Jumblies live were a good deal farther and fewer than they are affirmed to be: going to sea in a sieve is a mildly engaging idea, I suppose, but not so much so after six long elaborate stanzas and eighteen references to the sieve. The four lines allowed the wise men of Gotham seem nearer the mark. At times the threshold of pain is reached:

> The Pobble swam fast and well,
> And when boats or ships came near him,
> He tinkledy-binkledy-winkled a bell. . . .

Once or twice, notably in 'The Owl and the Pussy-Cat', Lear managed to avoid his besetting temptation, but the result is too touching to be light.

I have included a handful of his limericks with reluctance. If such a thing as a high-verse limerick were possible, the form he preferred—last line ending same as first or occasionally second—might be defended. As things are it is best avoided.

VII

I described *vers de société* earlier as to some degree a continuation of satire. This move can be seen as already in progress by the time of Swift, who was certainly not writing satire in the normal use of the term, as his avoidance of the heroic couplet would be enough to suggest. Before I go any further, an attempt at definition seems called for. We are dealing with a kind of realistic verse that is close to some of the interests of the novel: men and women among their fellows, seen as members of a group or class in a way that emphasizes manners, social forms, amusements, fashion (from millinery to philosophy), topicality, even gossip, all these treated in a bright, perspicuous style.

It may be safer to leave Swift's verse on one side as *sui generis*; in the earlier nineteenth century we reach the great or main period of *vers de société* as it is commonly thought of. Most of it, like most of the work of Tom Moore, too facile a writer for his own posthumous good, is founded on what was once the talk of the town and very soon thereafter ceased to be. Moore's friend Byron comes in here as well as anywhere, though he was doing several things at once, one of which was to bend social verse back towards satire—all the way in *The Vision of Judgment*. Or such was the intention: the performance bears out my early dating of the demise of satire by being most successful precisely where it is least satirical. Its movement carries the wrong kind of self-assurance, a fault Byron remedied in *Don Juan* by adhering to the manner of *Beppo*. His was the most flexible and expressive style ever devised for light verse. (Perhaps he had learnt something from his near-namesake, the mystic, shorthand pioneer and neglected light versifier John Byrom, whose poems went into a new edition—in Leeds, though—in 1814 after being out of print for decades. *Beppo* appeared in 1818.)

The social verse of that time often goes like this:

> Oh, where are the dandies who flirted,
> Who came of a morning to call?
> We females are so disconcerted,
> I'd *fee* males to come to my ball!
> 'Twas flattery charmed us—no matter—
> Paste often may pass for a gem;
> Alas! we are duller and flatter
> Than when we were flattered by them.

I quote this stanza by T. H. Bayly (1797–1839) partly to underline the greatness of his younger contemporary, W. M. Praed. Too interested in his material to have time for factitious word-play, Praed combines the smoothest possible texture with a full measure of the subversive quality I noted some pages back as a light-verse characteristic: a unique and deadly mixture. Not that Bayly is so bad, let it be said, and it is very likely to be said after a glance at the work of most poets born in this century.

Sir John Betjeman, Auden and Mr. Philip Larkin are notable exceptions. I realize that each of the three is up to something else besides writing *vers de société.*

VIII

In the arts that can be performed, very short first-rate pieces are rare. The only musical ones I can think of are Chopin's A major Prelude and the Last Post. There are some good 'high' epitaphs, but the verse epigram, likely in the nature of things to be light verse, requires caution on the anthologist's part as well as the poet's. Incredibly bad epigrams show an equally incredible power of being collected and re-collected over and over again. Turning the radio on at random the other week, I was horrified to hear someone saying (and not in tones of disgust, either):

> Swans sing before they die—'twere no bad thing
> Should certain persons die before they sing.

Take that! One feels that a really helpful editor should make quite sure no one could miss the point: '. . . *die* before they *sing*!!'—got it? This was Coleridge at play, but it was no more a person than Abel Evans (1679–1737) who wrote:

Tadlow

> When Tadlow walks the streets the paviours cry,
> 'God bless you, sir!' and lay their rammers by.

The first time I came across that, I deduced straight away that Tadlow must have been very fat indeed. On looking into the matter, I established that he was an Oxford don who flourished about 1700 and was very fat indeed. The Oxford don detail suggests to me an after-Hall common-room competition of the period requiring an impromptu on Tadlow to be composed within something like a minute. I should be quite tickled to have won with the above, but I hope I should have had the sense to throw it into the fire before staggering off to bed.

My search for the worst epigram in the language produced a strong field. Tom Moore was well up in front with the following (length of title in proportion to text is a primary point):

Impromptu

on being obliged to leave a pleasant party, from the want of a pair of breeches to dress for dinner in

Between Adam and me the great difference is,
Though a paradise each has been forced to resign,
That he never wore breeches, till turned out of his,
While, for want of my breeches, I'm banished from mine.

He ... *his* ... *want* ... *my* ... *I'm* ... *mine*, you see. The prize, however, goes to Samuel Rogers for this:

On a Certain Count Going to Italy and Leaving an Opera Score Behind Him

He has quitted the Countess—what can she wish more?
She loses one husband, and gets back a score.

There must be a stong consumer-demand for epigrams. I have tried to put up a measure of resistance.

It may be a too-conscious desire for pithiness and memorability, a will to wit, that trips up the would-be epigrammatist. The writers of what are usually regarded and collected as nursery rhymes (it must have been an advanced nursery in which, say, 'We're All Dry' appeared) seem more spontaneous, less tied down by form. From different periods and varied sources, a body of genuine not-quite-so-short light verse has emerged; now and then ('Trip upon Trenchers') it can move from the buoyant to the affecting with remarkable speed.

IX

There remains the remainder. I need hardly go so far as to mention other kinds of light verse. The comic song (Planché, 'She Was Poor but She Was Honest', Noël Coward) is not often successful on paper (and Coward at least was not exactly singing). The wit of the witty lyric as written by Cole Porter evaporates as it comes out of the end of the transcriber's pen. What, by musical analogy, can be called the verse study is found in Mr. John Fuller's 'Blues' in an unusually pure and concentrated form; 'An Austrian Army', 'The Vites' Journey' and 'I Saw a Fishpond' set their authors easier tasks, but there are strong elements of the same exercise—the graceful surmounting of self-imposed technical obstacles—in some of the best limericks. The comic poem, as written by Calverley, Gilbert and others of lesser note, needs no exposition. I have also included pieces that belong in more than one category or in none that I have distinguished.

I have not included any examples of unconscious humour in the style of William MacGonagall, the Rev. Cornelius Whur and such. The smile they may arouse is not ill-natured, to be sure, but after a short interval of fixity it will usually disappear altogether. Such authors hardly bear one reading, let alone more. And they were not writing light verse.

X

The reader will see that my selection comes nearly to an end with writers born in the decade of my own birth, the 1920s. This is not the result of any policy. Perhaps my taste is at fault, but the whole of this volume is at the mercy of that taste, and I must be allowed to suggest that what has limited me in my choice of recent verse is something that has happened to recent verse. I offer an example in no spirit of acrimony or mockery, but because it is the work of an established poet, is representative and is recent enough:

When all these men and
women came, in
the sunlight, to that
            tower they
found it
was embedded
in the earth. And
to get inside, you
crossed over this
iron bridge, to meet
spiralling
            downward
steps;
which they did,
and proceeded down
-stairs to a room
with only a
white
telephone in it and one
window looking out at hills
barely holding back the sea.

More follows in the same strain. It is no part of my commission to say that this is actually not verse at all in any sense that makes sense, though I will say so. What does concern me here is that when what is presumably aspiring to be high verse abandons form, a mortal blow is dealt to light verse, to which form has always been of the essence.

(Apparent exceptions like the performances of Mr. Tomlinson and Mr MacBeth—I am just about sure that the latter's are not verse, but have put them in to be on the unsafe side—are happy freaks.) Not merely can writings like the one quoted above not be parodied, they fail to provide that something to push against that light verse needs. You can't rhyme and scan with reference to, in relation to, alongside what does neither.

Light verse in the late 1970s consists almost entirely of political ephemera, often written with great skill and force; exercises in the styles of the dead or the ageing, for the most part in competitions in weekly journals; and limericks. I cannot see the situation improving much.

1977 KINGSLEY AMIS

CONTENTS

CONTENTS

CONTENTS

WILLIAM SHAKESPEARE

1564–1616

1 *Song: 'The master, the swabber'*

THE master, the swabber, the boatswain and I,
  The gunner and his mate,
Loved Mall, Meg, and Marian and Margery,
  But none of us cared for Kate;
  For she had a tongue with a tang,
  Would cry to a sailor, 'Go hang!'
She loved not the savour of tar nor of pitch,
Yet a tailor might scratch her where'er she did itch:
  Then to sea, boys, and let her go hang.

BEN JONSON

1573–1637

2 *On Giles and Joan*

WHO says that Giles and Joan at discord be?
The observing neighbours no such mood can see.
Indeed, poor Giles repents he married ever:
But that his Joan doth too. And Giles would never,
By his free will, be in Joan's company;
No more would Joan he should. Giles riseth early,
And having got him out of doors is glad—
The like is Joan—but turning home, is sad;
And so is Joan. Oft-times, when Giles doth find
Harsh sights at home, Giles wisheth he were blind:
All this doth Joan. Or that his long-yarned life
Were quite out-spun; the like wish hath his wife.
The children that he keeps, Giles swears are none
Of his begetting; and so swears his Joan:
In all affections she concurreth still.
If now, with man and wife, to will and nill
The self-same things a note of concord be,
I know no couple better can agree!

ANONYMOUS

3

'Oh that my lungs'

OH that my lungs could bleat like buttered peas;
  But bleating of my lungs hath caught the itch,
And are as mangy as the Irish seas
  That doth engender windmills on a bitch.

I grant that rainbows being lulled asleep,
  Snort like a woodknife in a lady's eyes;
Which makes her grieve to see a pudding creep,
  For creeping puddings only please the wise.

Not that a hard-roed herring should presume
  To swing a tithe-pig in a catskin purse;
For fear the hailstones which did fall at Rome,
  By lessening of the fault should make it worse.

For 'tis most certain winter woolsacks grow
  From geese to swans if men could keep them so,
Till that the sheep-shorn planets gave the hint
  To pickle pancakes in Geneva print.

Some men there were that did suppose the sky
  Was made of carbonadoed antidotes;
But my opinion is, a whale's left eye
  Need not be coined all King Harry groats.

The reason's plain, for Charon's western barge
  Running a tilt at the subjunctive mood,
Beckoned to Bednal Green, and gave him charge
  To fasten padlocks with Antarctic food.

The end will be the millponds must be laded,
  To fish for white pots in a country dance;
So they that suffered wrong and were upbraided
  Shall be made friends in a left-handed trance.

4

Hye Nonny Nonny Noe

DOWN lay the shepherd swain
  so sober and demure
Wishing for his wench again
  so bonny and so pure
With his head on hillock low
  and his arms akimbo,
And all was for the loss of his
  hye nonny nonny noe.

His tears fell as thin
  as water from the still,
His hair upon his chin
  grew like thyme upon a hill,
His cherry cheeks pale as snow
  did testify his mickle woe
And all was for the loss of his
  hye nonny nonny noe.

Sweet she was, as kind a love
  as ever fettered swain;
Never such a dainty one
  shall man enjoy again.
Set a thousand on a row
  I forbid that any show
Ever the like of her
  hye nonny nonny noe.

Face she had of filbert hue
  and bosomed like a swan
Back she had of bended ewe,
  and waisted by a span.
Hair she had as black as crow
  from the head unto the toe
Down down all over her
  hye nonny nonny noe.

With her mantle tucked up high
  she foddered her flock
So buxom and alluringly
  her knee upheld her smock

So nimbly did she use to go
  so smooth she danced on tip-toe,
That all the men were fond of her
  hye nonny nonny noe.

She smiled like a holy-day,
  she simpered like the spring
She pranked it like a popinjay,
  and like a swallow sing:
She tripped it like a barren doe,
  she strutted like a gorcrow,
Which made the men so fond of her
  hye nonny nonny noe.

To sport it on the merry down
  to dance the lively hay;
To wrestle for a green gown
  in heat of all the day
Never would she say me no
  yet me thought I had though
Never enough of her
  hye nonny nonny noe.

But gone she is the prettiest lass
  that ever trod on plain.
Whatever hath betide of her
  blame not the shepherd swain
For why she was her own foe,
  and gave herself the overthrow
By being so frank of her
  hye nonny nonny noe.

SAMUEL BUTLER

1612–1680

5 FROM *Satire upon the Licentious Age of Charles II*

How silly were those sages heretofore
To fright their heroes with a syren-whore?
Make 'em believe a water-witch with charms
Could sink their men-of-war as easy as storms,

And turn their mariners, that heard them sing,
Into land-porpoises, and cod, and ling,
To terrify those mighty champions,
As we do children now with bloody-bones;
Until the subtlest of their conjurors
Sealed up the labels to his soul, his ears,
And tied his deafened sailors (while he passed
The dreadful lady's lodgings) to the mast,
And rather venture drowning than to wrong
The sea-pugs' chaste ears with a bawdy song:
To be out of countenance, and like an ass
Not pledge the lady Circe one beer-glass;
Unmannerly refuse her treat and wine
For fear of being turned into a swine;
When one of our heroic adventurers now
Would drink her down, and turn her into a sow.

6 FROM *Hudibras*

In mathematics he was greater
Than Tycho Brahe, or Erra Pater:
For he by geometric scale
Could take the size of pots of ale;
Resolve by sines and tangents straight
If bread or butter wanted weight;
And wisely tell what hour o' the day
The clock doth strike, by algebra.

Beside he was a shrewd philosopher,
And had read every text and gloss over:
Whate'er the crabbed'st author hath
He understood by implicit faith,
Whatever sceptic could inquire for;
For every why he had a wherefore:
Knew more than forty of them do,
As far as words and terms could go.
All which he understood by rote,
And as occasion served, would quote;
No matter whether right or wrong:
They might be either said or sung.
His notions fitted things so well,

That which was which he could not tell;
But oftentimes mistook the one
For the other, as great clerks have done.
He could reduce all things to acts,
And knew their natures by abstracts,
Where entity and quiddity,
The ghosts of defunct bodies, fly;
Where truth in person does appear,
Like words congealed in northern air.
He knew what's what, and that's as high
As metaphysic wit can fly,
In School-Divinity as able
As he that hight Irrefragable;
Profound in all the nominal
And real ways beyond them all,
And with as delicate a hand
Could twist as tough a rope of sand,
And weave fine cobwebs, fit for skull
That's empty when the moon is full;
Such as take lodgings in a head
That's to be let unfurnished.
He could raise scruples dark and nice,
And after solve 'em in a trice:
As if Divinity had catched
The itch, of purpose to be scratched;
Or, like a mountebank, did wound
And stab herself with doubts profound,
Only to show with how small pain
The sores of faith are cured again;
Although by woeful proof we find,
They always leave a scar behind.
He knew the seat of Paradise,
Could tell in what degree it lies:
And as he was disposed, could prove it,
Below the moon, or else above it:
What Adam dreamt of when his bride
Came from her closet in his side:
Whether the Devil tempted her
By a High Dutch interpreter:
If either of them had a navel;
Who first made music malleable:
Whether the serpent at the Fall
Had cloven feet, or none at all.

All this, without a gloss or comment,
He would unriddle in a moment
In proper terms, such as men smatter
When they throw out and miss the matter.

For his religion, it was fit
To match his learning and his wit:
'Twas Presbyterian true blue,
For he was of that stubborn crew
Of errant saints, whom all men grant
To be the true Church Militant:
Such as do build their faith upon
The holy Text of pike and gun;
Decide all controversies by
Infallible artillery;
And prove their doctrine orthodox
By apostolic blows and knocks;
Call fire and sword and desolation,
A godly-thorough-Reformation,
Which always must be carried on,
And still be doing, never done:
As if religion were intended
For nothing else but to be mended.
A sect, whose chief devotion lies
In odd perverse antipathies;
In falling out with that or this,
And finding somewhat still amiss:
More peevish, cross, and splenetic,
Than dog distract, or monkey sick.
That with more care keep holy-day
The wrong, than others the right way:
Compound for sins they are inclined to,
By damning those they have no mind to;
Still so perverse and opposite,
As if they worshipped God for spite,
The self-same thing they will abhor
One way, and long another for.
Free-will they one way disavow,
Another, nothing else allow.
All piety consists therein
In them, in other men all sin.
Rather than fail, they will defy
That which they love most tenderly,

Quarrel with minced pies, and disparage
Their best and dearest friend, plum-porridge;
Fat pig and goose itself oppose,
And blaspheme custard through the nose.
The apostles of this fierce religion,
Like Mahomet's, were ass and widgeon,
To whom our Knight, by fast instinct
Of wit and temper was so linked,
As if hypocrisy and nonsense
Had got the advowson of his conscience.

*

The question then, to state it first,
Is which is better, or which worst,
Synods or bears. Bears I avow
To be the worst, and synods thou.
But to make good the assertion,
Thou sayst they're really all one.
If so, not worst; for if they're *idem*,
Why then, *tantundem dat tantidem.*
For if they are the same, by course
Neither is better, neither worse.
But I deny they are the same,
More than a maggot and I am.
That both are *animalia*,
I grant, but not *rationalia*:
For though they do agree in kind,
Specific difference we find.
And can no more make bears of these
Than prove my horse is Socrates.

That synods are bear-gardens too,
Thou dost affirm; but I say no:
And thus I prove it, in a word,
Whatsoever assembly's not empowered
To censure, curse, absolve, and ordain,
Can be no synod: but bear-garden
Has no such power, *ergo* 'tis none.
And so thy sophistry's o'erthrown.

But yet we are beside the question
Which thou didst raise the first contest on;
For that was, whether bears are better
Than synod-men; I say, *negatur*.

That bears are beasts, and synods men,
Is held by all: they're better then.
For bears and dogs on four legs go,
As beasts, but synod-men on two.
'Tis true, they all have teeth and nails;
But prove that synod-men have tails,
Or that a rugged, shaggy fur
Grows o'er the hide of Presbyter,
Or that his snout and spacious ears
Do hold proportion with a bear's.
A bear's a savage beast, of all
Most ugly and unnatural,
Whelped without form, until the dam
Have licked him into shape and frame:
But all thy light can ne'er evict
That ever synod-man was licked.

*

This place (quoth she) they say's enchanted,
And with delinquent spirits haunted;
That here are tied in chains, and scourged,
Until their guilty crimes be purged;
Look, there are two of them appear
Like persons I have seen somewhere:
Some have mistaken blocks and posts,
For spectres, apparitions, ghosts
With saucer-eyes and horns; and some
Have heard the Devil beat a drum:
But if our eyes are not false glasses,
That give a wrong account of faces,
That beard and I should be acquainted
Before 'twas conjured and enchanted.
For though it be disfigured somewhat,
As if it had lately been in combat,
It did belong to a worthy Knight,
Howe'er this goblin is come by it.

When Hudibras the lady heard
To take kind notice of his beard,
And speak with such respect and honour,
Both of the beard, and the beard's owner,
He thought it best to set as good
A face upon it as he could,

And thus he spoke; Lady, your bright
And radiant eyes are in the right:
The beard's the identic beard you knew,
The same numerically true:
Nor is it worn by fiend or elf,
But its proprietor himself.

Oh Heavens! quoth she, can that be true?
I do begin to fear 'tis you:
Not by your individual whiskers,
But by your dialect and discourse;
That never spoke to man or beast
In notions vulgarly expressed.
But what malignant star, alas,
Has brought you both to this sad pass?
Quoth he, the fortune of the war,
Which I am less afflicted for,
Than to be seen with beard and face,
By you in such a homely case.
Quoth she, those need not be ashamed
For being honourably maimed;
If he that is in battle conquered
Have any title to his own beard,
Though yours be sorely lugged and torn,
It does your visage more adorn,
Than if 'twere pruned, and starched, and laundered,
And cut square by the Russian standard.
A torn beard's like a tattered ensign,
That's bravest which there are most rents in.

*

Quoth he, to bid me not to love,
Is to forbid my pulse to move,
My beard to grow, my ears to prick up,
Or (when I'm in a fit) to hiccup;
Command me to piss out the moon
And 'twill as easily be done.
Love's power's too great to be withstood,
By feeble human flesh and blood.
'Twas he that brought upon his knees
The hectoring kill-cow Hercules;
Reduce his leaguer-lion's skin
To a petticoat, and made him spin:

Seized on his club, and made it dwindle
To a feeble distaff, and a spindle.
'Twas he made Emperors gallants
To their own sisters, and their aunts;
Set Popes and Cardinals agog,
To play with pages at leap-frog;
'Twas he, that gave our senate purges,
And fluxed the House of many a burgess;
Made those that represent the nation
Submit, and suffer amputation:
And all the grandees o' the Cabal,
Adjourn to tubs at spring and fall.
He mounted synod-men and rode 'em
To Dirty Lane, and Little Sodom;
Made 'em curvet, like Spanish jennets,
And take the ring at Madam——.
'Twas he that made Saint Francis do
More than the Devil could tempt him to;
In cold and frosty weather, grow
Enamoured of a wife of snow;
And though she were of rigid temper,
With melting flames accost and tempt her:
Which after in enjoyment quenching,
He hung a garland on his engine.

*

What makes a knave a child of God,
And one of us?—A livelihood.
What renders beating-out of brains
And murder godliness?—Great gains.
What makes you encroach upon our trade,
And damn all others?—To be paid.
What's orthodox, and true believing
Against a conscience?—A good living.
What makes rebelling against Kings
A Good Old Cause? Administrings.
What makes all doctines plain and clear?
About two hundred pounds a year.
And that which was proved true before,
Prove false again? Two hundred more.
What makes the breaking of all oaths
A holy duty? Food and clothes.

What laws, and freedom, persecution?
Being out of power, and contribution.
What makes a church a den of thieves?
A Dean, a Chapter, and white sleeves.
And what would serve if those were gone,
To make it orthodox? Our own . . .
What's liberty of conscience,
In the natural and genuine sense?
'Tis to restore with more security
Rebellion to its ancient purity.

ANDREW MARVELL

1621–1678

7 *The Character of Holland*

HOLLAND, that scarce deserves the name of land,
As but the off-scouring of the British sand,
And so much earth as was contributed
By English pilots when they heaved the lead,
Or what by the ocean's slow alluvion fell,
Of shipwrecked cockle and the mussel-shell;
This indigested vomit of the sea
Fell to the Dutch by just propriety.
  Glad then, as miners that have found the ore,
They with mad labour fished the land to shore,
And dived as desperately for each piece
Of earth, as if't had been of ambergris;
Collecting anxiously small loads of clay,
Less than what building swallows bear away;
Or than those pills which sordid beetles roll,
Transfusing into them their dunghill soul.
  How did they rivet, with gigantic piles,
Thorough the centre their new-catched miles,
And to the stake a struggling country bound,
Where barking waves still bait the forced ground;
Building their watery Babel far more high
To reach the sea, then those to scale the sky.

Yet still his claim the injured ocean laid,
And oft at leap-frog o'er their steeples played:
As if on purpose it on land had come
To show them what's their *Mare Liberum.*
A daily deluge over them does boil;
The earth and water play at level-coil;
The fish oft-times the burgher dispossessed,
And sat not as a meat but as a guest;
And oft the tritons and the sea-nymphs saw
Whole shoals of Dutch served up for Cabilliau;
Or as they over the new level ranged
For pickled herring, pickled Heeren changed.
Nature, it seemed, ashamed of her mistake,
Would throw their land away at duck and drake.
Therefore Necessity, that first made Kings,
Something like Government among them brings.
For as with pygmies who best kills the crane,
Among the hungry he that treasures grain,
Among the blind the one-eyed blinkard reigns,
So rules among the drowned he that drains.
Not who first see the rising sun commands,
But who could first discern the rising lands.
Who best could know to pump an earth so leak
Him they their lord and country's father speak.
To make a bank was a great plot of state;
Invent a shovel and be a magistrate.
Hence some small Dyke-grave unperceived invades
The power, and grows as 'twere a King of Spades.
But for less envy some joint states endures,
Who look like a commission of the sewers.
For these Half-anders, half wet and half dry,
Nor bear strict service, nor pure liberty.
'Tis probable Religion after this
Came next in order; which they could not miss
How could the Dutch but be converted, when
The Apostles were so many fishermen?
Besides the waters of themselves did rise,
And, as their land, so them did re-baptize.
Though herring for their God few voices missed,
And Poor-John to have been the Evangelist,
Faith, that could never twins conceive before,
Never so fertile, spawned upon this shore:
More pregnant then their Margaret, that laid down

For Hans-in-Kelder of a whole Hans-Town.
  Sure when Religion did itself embark,
And from the east would westward steer its ark,
It struck, and splitting on this unknown ground,
Each one thence pillaged the first piece he found:
Hence Amsterdam, Turk-Christian-Pagan-Jew,
Staple of sects and mint of schism grew;
That bank of conscience, where not one so strange
Opinion but finds credit and exchange.
In vain for Catholics ourselves we bear;
The universal Church is only there.
Nor can Civility there want for tillage,
Where wisely for their court they chose a village.
How fit a title cloths their Governors,
Themselves the hogs as all their subjects boars!
  Let it suffice to give their country fame
That it had one Civilis called by name
Some fifteen hundred and more years ago,
But surely never any that was so.
  See but their mermaids with their tails of fish,
Reeking at church over the chafing-dish.
A vestal turf enshrined in earthenware
Fumes through the loop-holes of a wooden square.
Each to the temple with these altars tend,
But still does place it at her western end,
While the fat steam of female sacrifice
Fills the priest's nostrils and puts out his eyes. . . .
  But when such amity at home is showed,
What then are their confederacies abroad?
Let this one courtesy witness all the rest;
When their whole navy they together pressed,
Not Christian captives to redeem from bands,
Or intercept the western golden sands,
No, but all ancient rights and leagues must vail,
Rather than to the English strike their sail;
To whom their weather-beaten province owes
Itself, when as some greater vessel tows
A cock-boat tossed with the same wind and fate;
We buoyed so often up their sinking state.
  Was this *Jus Belli & Pacis*? Could this be
Cause why their burgomaster of the sea
Rammed with gunpowder, flaming with brand-wine,
Should raging hold his linstock to the mine?

While, with feigned treaties, they invade by stealth
Our sore new circumcised Commonwealth.
  Yet of his vain attempt no more he sees
Then of case-butter shot and bullet-cheese;
And the torn navy staggered with him home,
While the sea laughed itself into a foam,
'Tis true since that (as fortune kindly sports)
A wholesome danger drove us to our ports,
While half their banished keels the tempest tossed,
Half bound at home in prison to the frost:
That ours meantime at leisure might careen,
In a calm winter, under skies serene.
As the obsequious air and waters rest,
Till the dear halcyon hatch out all its nest.
The Commonwealth doth by its losses grow;
And, like its own seas, only ebbs to flow.
Besides that very agitation laves,
And purges out the corruptible waves.
  And now again our armed *Bucentaur*
Doth yearly their sea-nuptials restore.
And now the hydra of seven provinces
Is strangled by our infant Hercules.
Their tortoise wants its vainly stretched neck,
Their navy all our conquest or our wreck:
Or, what is left, their Carthage overcome
Would render fain unto our better Rome. . . .
  For now of nothing may our state despair,
Darling of Heaven, and of men the care;
Provided that they be what they have been,
Watchful abroad, and honest still within.
For while our *Neptune* doth a trident shake,
Steeled with those piercing heads, Dean, Monck and
    Blake.
And while Jove governs in the highest sphere,
Vainly in hell let Pluto domineer.

ANONYMOUS

8 *Three Children*

THREE children sliding on the ice,
Upon a summer's day,
As it fell out, they all fell in,
The rest they ran away.

Now had these children been at home,
Or sliding on dry ground,
Ten thousand pounds to one penny
They had not all been drowned.

You parents all that children have,
And you that have got none,
If you would have them safe abroad,
Pray keep them safe at home.

9 *A Description of Maidenhead*

HAVE you not in a chimney seen
A sullen faggot, wet and green,
How coyly it receives the heat
And at both ends does fume and sweat?

So fares it with the harmless maid
When first upon her back she's laid;
But the kind experienced dame
Cracks, and rejoices in the flame.

THOMAS FLATMAN

1637–1688

10

On Marriage

How happy a thing were a wedding,
  And a bedding,
If a man might purchase a wife
  For a twelvemonth and a day;
But to live with her all a man's life,
  For ever and for aye,
Till she grow as grey as a cat,
Good faith, Mr. Parson, I thank you for that!

JOHN WILMOT, EARL OF ROCHESTER

1647–1680

11

Song

Love a woman? You're an ass!
  'Tis a most insipid passion
To choose out for your happiness
  The silliest part of God's creation.

Let the porter and the groom,
  Things designed for dirty slaves,
Drudge in fair Aurelia's womb
  To get supplies for age and graves.

Farewell, woman! I intend
  Henceforth every night to sit
With my lewd, well-natured friend,
  Drinking to engender wit.

Then give me health, wealth, mirth, and wine,
  And, if busy love entrenches,
There's a sweet, soft page of mine
  Does the trick worth forty wenches.

12 *The Disabled Debauchee*

As some brave admiral, in former war
  Deprived of force, but pressed with courage still,
Two rival fleets appearing from afar,
  Crawls to the top of an adjacent hill;

From whence, with thoughts full of concern, he views
  The wise and daring conduct of the fight,
Whilst each bold action to his mind renews
  His present glory and his past delight;

From his fierce eyes flashes of fire he throws,
  As from black clouds when lightning breaks away;
Transported, thinks himself amidst the foes,
  And absent, yet enjoys the bloody day;

So, when my days of impotence approach,
  And I'm by pox and wine's unlucky chance
Forced from the pleasing billows of debauch
  On the dull shore of lazy temperance,

My pains at least some respite shall afford
  While I behold the battles you maintain
When fleets of glasses sail about the board,
  From whose broadsides volleys of wit shall rain.

Nor let the sight of honourable scars,
  Which my too forward valour did procure,
Frighten new-listed soldiers from the wars:
  Past joys have more than paid what I endure.

Should any youth (worth being drunk) prove nice,
  And from his fair inviter meanly shrink,
'Twill please the ghost of my departed vice
  If, at my counsel, he repent and drink.

Or should some cold-complexioned sot forbid,
  With his dull morals, our bold night-alarms,
I'll fire his blood by telling what I did
  When I was strong and able to bear arms.

I'll tell of whores attacked, their lords at home;
  Bawds' quarters beaten up, and fortress won;
Windows demolished, watches overcome;
  And handsome ills by my contrivance done.

With tales like these I will such thoughts inspire
  As to important mischief shall incline:
I'll make him long some ancient church to fire,
  And fear no lewdness he's called to by wine.

Thus, statesmanlike, I'll saucily impose,
  And safe from action, valiantly advise;
Sheltered in impotence, urge you to blows,
  And being good for nothing else, be wise.

13

Impromptu on Charles II

GOD bless our good and gracious King,
  Whose promise none relies on;
Who never said a foolish thing,
  Nor ever did a wise one.

DANIEL DEFOE

1661?–1731

14

FROM *The True-Born Englishman*

WHEREVER God erects a house of prayer,
The Devil always builds a chapel there:
And 'twill be found upon examination
The latter has the largest congregation:
For ever since he first debauched the mind,
He made a perfect conquest of mankind.
With uniformity of service, he
Reigns with a general aristocracy.
No non-conforming sects disturb his reign,
For of his yoke, there's very few complain.

He knows the genius and the inclination,
And matches proper sins for every nation,
He needs no standing army government;
He always rules us by our own consent. . . .
  By zeal the Irish, and the Russ by folly,
Fury the Dane, the Swede by melancholy;
By stupid ignorance, the Muscovite;
The Chinese, by a child of hell, called wit;
Wealth makes the Persian too effeminate;
And poverty the Tartar desperate:
The Turks and Moors, by Mah'met he subdues;
And God has given him leave to rule the Jews;
Rage rules the Portuguese, and fraud the Scotch;
Revenge the Pole, and avarice the Dutch.
  Satire, be kind, and draw a silent veil,
Thy native England's vices to conceal:
Or, if that task's impossible to do,
At least be just, and show her virtues too;
Too great the first, alas! the last too few.
  England, unknown, as yet, unpeopled lay,—
Happy, had she remained so to this day,
And not to every nation been a prey. . . .
  The Romans first with Julius Cæsar came,
Including all the nations of that name,
Gauls, Greeks, and Lombards; and by computation,
Auxiliaries or slaves of every nation.
With Hengist, Saxons; Danes with Sweno came,
In search of plunder, not in search of fame.
Scots, Picts, and Irish from the Hibernian shore;
And conquering William brought the Normans o'er.
  All these their barbarous offspring left behind,
The dregs of armies, they of all mankind;
Blended with Britons, who before were here,
Of whom the Welsh have blessed the character.
  Thus from a mixture of all kinds began,
That heterogeneous thing, an Englishman:
In eager rapes, and furious lust begot,
Betwixt a painted Briton and a Scot:
Whose gendering offspring quickly learned to bow,
And yoke their heifers to the Roman plough;
From whence a mongrel half-bred race there came,
With neither name nor nation, speech nor fame,
In whose hot veins new mixtures quickly ran,

Infused betwixt a Saxon and a Dane;
While their rank daughters, to their parents just,
Received all nations with promiscuous lust.
This nauseous brood directly did contain
The well-extracted blood of Englishmen. . . .
The Scot, Pict, Briton, Roman, Dane submit,
And with the English Saxon all unite:
And these the mixture have so close pursued,
The very name and memory's subdued;
No Roman now, no Briton, does remain;
Wales strove to separate, but strove in vain:
The silent nations undistinguished fall,
And Englishman's the common name for all.
Fate jumbled them together, God knows how;
Whate'er they were, they're true-born English now.

*

The labouring poor, in spite of double pay,
Are saucy, mutinous, and beggarly;
So lavish of their money and their time,
That want of forecast is the nation's crime.
Good drunken company is their delight;
And what they get by day they spend by night.
Dull thinking seldom does their heads engage,
But drink their youth away, and hurry on old age.
Empty of all good husbandry and sense;
And void of manners most when void of pence.
Their strong aversion to behaviour's such,
They always talk too little or too much.
So dull, they never take the pains to think;
And seldom are good natured but in drink. . . .
Nor do the poor alone their liquor prize,
The sages join in this great sacrifice;
The learned men who study Aristotle,
Correct him with an explanation bottle. . . .
The doctors too their Galen here resign,
And generally prescribe specific wine;
The graduate's study's grown an easy task,
While for the urinal they toss the flask;
The surgeon's art grows plainer every hour,
And wine's the balm which into wounds they pour. . . .
Statesmen their weighty politics refine,
And soldiers raise their courages by wine. . . .

Some think the clergy first found out the way,
And wine's the only spirit by which they pray.
But others, less profane than so, agree,
It clears the lungs, and helps the memory:
And, therefore, all of them divinely think,
Instead of study, 'tis as well to drink.

SAMUEL WESLEY

1662–1735

15 *A Pindaric on the Grunting of a Hog*

FREEBORN Pindaric never does refuse
Either a lofty or a humble Muse:
Now in proud Sophoclean buskins sings
Of heroes and of kings,
Mighty numbers, mighty things;
Now out of sight she flies,
Rowing with gaudy wings
Across the stormy skies;
Then down again
Herself she flings,
Without uneasiness or pain,
To lice and dogs
To cows and hogs,
And follows their melodious grunting o'er the plain.

Harmonious hog, draw near!
No bloody butcher's here,
Thou needst not fear.
Harmonious hog, draw near, and from thy beauteous
snout
(Whilst we attend with ear,
Like thine, pricked up devout,
To taste thy sugary voice, which here and there,
With wanton curls, vibrates around the circling
air),
Harmonious hog! warble some anthem out!
As sweet as those which quavering monks, in days
of yore,
With us did roar,
When they (alas

That the hard-hearted abbot such a coil should keep,
And cheat 'em of their first, their sweetest sleep!)
When they were ferreted up to midnight mass:
Why should not the pigs on organs play,
As well as they?
Dear hog! thou king of meat!
So near thy lord, mankind,
The nicest taste can scarce a difference find!
No more may I thy glorious gammons eat—
No more
Partake of the free farmer's Christmas store,
Black puddings which with fat would make your
mouth run o'er—
If I (though I should ne'er so long the sentence
stay,
And in my large ears' scale the thing ne'er so
discreetly weigh),
If I can find a difference in the notes
Belched from the applauded throats
Of rotten playhouse songsters all divine,—
If any difference I can find between their notes and
thine,
A noise they keep, with tune and out of tune,
And round and flat,
High, low, and this and that,
That Algebra or thou or I might understand
as soon.

Like the confounding lute's innumerable strings
One of them sings.
Thy easier music's ten times more divine:
More like the one-stringed, deep, majestic trump-
marine.
Prithee strike up, and cheer this drooping heart of
mine.
Not the sweet harp that's claimed by Jews,
Nor that which to the far more ancient Welsh
belongs,
Nor that which the wild Irish use,
Frighting even their own wolves with loud hub-
bubbaboos,
Nor Indian dance with Indian songs,
Nor yet

(Which how should I so long forget?)
The crown of all the rest,
The very cream o' the jest,
Amphion's noble lyre—the tongs;
Nor, though poetic Jordan bite his thumbs
At the bold world, my Lord Mayor's flutes and kettledrums;
Not all this instrumental dare
With thy soft, ravishing, vocal music ever to compare!

WILLIAM WALSH

1663–1708

16

The Despairing Lover

DISTRACTED with care,
For Phyllis the fair;
Since nothing could move her,
Poor Damon, her lover,
Resolves in despair
No longer to languish,
Nor bear so much anguish;
But, mad with his love,
To a precipice goes;
Where, a leap from above
Would soon finish his woes.
When in rage he came there,
Beholding how steep
The sides did appear,
And the bottom how deep;
His torments projecting,
And sadly reflecting,
That a lover forsaken
A new love may get;
But a neck, when once broken,
Can never be set:
And, that he could die
Whenever he would;

But, that he could live
But as long as he could;
How grievous soever
The torment might grow,
He scorned to endeavour
To finish it so.
But bold, unconcerned
At the thoughts of the pain,
He calmly returned
To his cottage again.

JONATHAN SWIFT

1667–1745

17

Daphne

DAPHNE knows, with equal ease,
How to vex and how to please,
But the folly of her sex
Makes her sole delight to vex.
Never woman more devised
Surer ways to be despised:
Paradoxes weakly wielding,
Always conquered, never yielding.
To dispute, her chief delight,
With not one opinion right:
Thick her arguments she lays on,
And with cavils combats reason:
Answers in decisive way,
Never hears what you can say:
Still her odd perverseness shows
Chiefly where she nothing knows.
And where she is most familiar,
Always peevisher and sillier:
All her spirits in a flame
When she knows she's most to blame.

Send me hence ten thousand miles
From a face that always smiles:
None could ever act that part
But a Fury in her heart.

Ye who hate such inconsistence,
To be easy keep your distance;
Or in folly still befriend her,
But have no concern to mend her.
Lose not time to contradict her,
Nor endeavour to convict her.
Never take it in your thought
That she'll own or cure a fault.
Into contradiction warm her,
Then, perhaps, you may reform her:
Only take this rule along,
Always to advise her wrong;
And reprove her when she's right;
She may then grow wise for spite.

No—that scheme will ne'er succeed,
She has better learnt her creed:
She's too cunning and too skilful,
When to yield and when be willful.
Nature holds her forth two mirrors,
One for truth and one for errors:
That looks hideous, fierce and frightful;
This is flattering and delightful;
That she throws away as foul;
Sits by this to dress her soul.

Thus you have the case in view,
Daphne, 'twixt the Dean and you,
Heaven forbid he should despise thee;
But will never more advise thee.

Twelve Articles

1. Lest it may more quarrels breed
   I will never hear you read.
2. By disputing I will never
   To convince you, once endeavour.
3. When a paradox you stick to,
   I will never contradict you.
4. When I talk, and you are heedless,
   I will show no anger needless.
5. When your speeches are absurd,
   I will ne'er object a word.

6. When you furious argue wrong,
   I will grieve, and hold my tongue.
7. Not a jest or humorous story
   Will I ever tell before ye:
   To be chidden for explaining
   When you quite mistake the meaning.
8. Never more will I suppose
   You can taste my verse or prose:
9. You no more at me shall fret,
   While I teach, and you forget;
10. You shall never hear me thunder,
    When you blunder on, and blunder.
11. Show your poverty of spirit,
    And in dress place all your merit;
    Give yourself ten thousand airs;
    That with me shall break no squares.
12. Never will I give advice
    Till you please to ask me thrice;
    Which if you in scorn reject,
    'Twill be just as I expect.

Thus we both shall have our ends,
And continue special friends.

18 FROM *The Life and Character of Dean Swift*

WISE Rochefoucault a maxim writ,
Made up of malice, truth, and wit:
If what he says be not a joke,
We mortals are strange kind of folk.
But hold: before we farther go,
'Tis fit the maxim we should know.

He says, 'Whenever Fortune sends
Disasters to our dearest friends,
Although we outwardly may grieve,
We oft are laughing in our sleeve.'
And, when I think upon't, this minute,
I fancy there is something in it.

We see a comrade get a fall,
Yet laugh our hearts out, one and all.

Tom for a wealthy wife looks round,
A nymph that brings ten thousand pound:
He nowhere could have better picked;
A rival comes, and Tom—is nicked.
See, how behave his friends professed,
They turn the matter to a jest;
Loll out their tongues, and thus they talk,
Poor Tom has got a plaguey baulk!

I could give instances enough,
That human friendship is but stuff.
Whene'er a flattering puppy cries
You are his dearest friend, he lies:
To lose a guinea at picquet,
Would make him rage and storm and fret,
Bring from his heart sincerer groans
Than if he heard you broke your bones.

Come, tell me truly, would you take well,
Suppose your friend and you were Equal,
To see him always foremost stand,
Affect to take the upper hand,
And strive to pass, in public view,
For much a better man than you?
Envy, I doubt, would powerful prove,
And get the better of your love;
'Twould please your palate, like a feast,
To see him mortified at least.

'Tis true, we talk of friendship much,
But who are they that can keep touch?
True friendship in two breasts requires
The same aversions and desires;
My friend should have, when I complain,
A fellow-feeling of my pain.

Yet, by experience, oft we find
Our friends are of a different mind;
And, were I tortured with the gout,
They'd laugh to see me make a rout,
Glad that themselves could walk about.

JONATHAN SWIFT

Let me suppose two special friends,
And each to poetry pretends:
Would either poet take it well
To hear the other bore the bell?
His rival for the chiefest reckoned,
Himself pass only for the second?

When you are sick, your friends, you say,
Will send their Howd'y's every day:
Alas! that gives you small relief!
They send for manners, not for grief:
Nor, if you died, would fail to go
That evening to a puppet show:
Yet, come in time to show their loves,
And get a hatband, scarf and gloves.

To make these truths the better known,
Let me suppose the case my own.

The day will come, when't shall be said,
'D'ye hear the news? The Dean is dead!
Poor man! he went all on a sudden!'
He's dropped, and given the crow a pudden!
What money was behind him found?
'I hear about two thousand pound.
'Tis owned he was a man of wit.'
Yet many a foolish thing he writ;
'And, sure he must be deeply learned—!'
That's more than ever I discerned;
'I know his nearest friends complain,
He was too airy for a Dean.
He was an honest man I'll swear.'
Why sir, I differ from you there,
For I have heard another story:
He was a most confounded Tory!
'Yet here we had a strong report,
That he was well-received at Court.'
Why, then it was, I do assert,
Their goodness, more than his desert.
He grew, or else his comrades lied,
Confounded dull before he died.

19 FROM *Verses on the Death of Doctor Swift*

THE time is not remote, when I
Must by the course of nature die:
When I foresee my special friends
Will try to find their private ends:
Though it is hardly understood
Which way my death can do them good;
Yet thus, methinks, I hear 'em speak;
'See, how the Dean begins to break,
Poor Gentleman, he droops apace,
You plainly find it in his face;
That old vertigo in his head
Will never leave him till he's dead;
Besides, his memory decays,
He recollects not what he says,
He cannot call his friends to mind,
Forgets the place where last he dined,
Plies you with stories o'er and o'er—
He told them fifty times before.
How does he fancy we can sit,
To hear his out-of-fashioned wit?
But he takes up with younger folks,
Who for his wine will bear his jokes:
Faith, he must make his stories shorter,
Or change his comrades once a quarter;
In half the time, he talks them round,
There must another set be found.

For poetry, he's past his prime,
He takes an hour to find a rhyme;
His fire is out, his wit decayed,
His fancy sunk, his muse a jade.
I'd have him throw away his pen,
But there's no talking to some men.

And then their tenderness appears,
By adding largely to my years:
'He's older than he would be reckoned,
And well remembers Charles the Second.

'He hardly drinks a pint of wine,
And that, I doubt, is no good sign.

His stomach too begins to fail:
Last year we thought him strong and hale,
But now, he's quite another thing;
I wish he may hold out till spring.'

Then hug themselves, and reason thus;
'It is not yet so bad with us.'

In such a case they talk in tropes,
And by their fears express their hopes:
Some great misfortune to portend,
No enemy can match a friend;
With all the kindness they profess,
The merit of a lucky guess,
(When daily Howd'y's come of course,
And servants answer, worse and worse)
Would please 'em better than to tell
That, God be praised, the Dean is well.
Then he who prophesied the best
Approves his foresight to the rest:
'You know, I always feared the worst,
And often told you so at first':
He'd rather choose that I should die
Than his prediction prove a lie.
Not one foretells I shall recover,
But all agree to give me over.

Yet should some neighbour feel a pain,
Just in the parts where I complain,
How many a message would he send?
What hearty prayers that I should mend?
Enquire what regimen I kept,
What gave me ease, and how I slept?
And more lament, when I was dead,
Than all the snivellers round my bed.

My good companions, never fear,
For though you may mistake a year;
Though your prognostics run too fast,
They must be verified at last. . . .

The doctors tender of their fame,
Wisely on me lay all the blame:

'We must confess his case was nice,
But he would never take advice:
Had he been ruled, for ought appears,
He might have lived these twenty years:
For when we opened him we found,
That all his vital parts were sound.'

From Dublin soon to London spread,
'Tis told at Court, the Dean is dead. . . .

My female friends, whose tender hearts
Have better learned to act their parts,
Receive the news in doleful dumps,
'The Dean is dead' (and what is trumps?)
Then Lord have mercy on his soul
(Ladies, I'll venture for the Vole).
'Six Deans they say must bear the pall
(I wish I knew what king to call).
Madam, your husband will attend
The funeral of so good a friend?'
No madam, 'tis a shocking sight,
And he's engaged Tomorrow night!
My lady Club would take it ill,
If he should fail her at quadrille!
He loved the Dean (I lead a heart),
But dearest friends, they say, must part.
His time was come, he ran his race;
We hope he's in a better place.

Why do we grieve that friends should die?
No loss more easy to supply.
One year is past; a different scene;
No further mention of the Dean,
Who now, alas, no more is missed
Than if he never did exist.
Where's now this favourite of Apollo?
Departed, and his works must follow:
Must undergo the common fate;
His kind of wit is out of date.

JOHN GAY

1685–1732

20

A New Song of New Similes

My passion is as mustard strong;
  I sit, all sober sad;
Drunk as a piper all day long,
  Or like a March hare mad.

Round as a hoop the bumpers flow,
  I drink, yet can't forget her;
For tho' as drunk as David's sow,
  I love her still the better.

Pert as a pear-monger I'd be,
  If Molly were but kind;
Cool as a cucumber could see
  The rest of womankind.

Like a stuck pig I gaping stare,
  And eye her o'er and o'er;
Lean as a rake with sighs and care,
  Sleek as a mouse before.

Plump as a partridge was I known,
  And soft as silk my skin,
My cheeks as fat as butter grown;
  But as a groat now thin!

I melancholy as a cat,
  Am kept awake to weep;
But she insensible of that,
  Sound as a top can sleep.

Hard is her heart as flint or stone,
  She laughs to see me pale,
And merry as a grig is grown,
  And brisk as bottled ale.

JOHN GAY

The God of Love at her approach
  Is busy as a bee,
Hearts sound as any bell or roach
  Are smit and sigh like me.

Ay me, as thick as hops or hail,
  The fine men crowd about her;
But soon as dead as a door-nail
  Shall I be if without her.

Strait as my leg her shape appears,
  O were we joined together!
My heart would be scot-free from cares,
  And lighter than a feather.

As fine as fivepence is her mien,
  No drum was ever tighter;
Her glance is as the razor keen,
  And not the sun is brighter.

As soft as pap her kisses are,
  Methinks I taste them yet.
Brown as a berry is her hair,
  Her eyes as black as jet;

As smooth as glass, as white as curds,
  Her pretty hand invites;
Sharp as a needle are her words,
  Her wit like pepper bites:

Brisk as a body-louse she trips,
  Clean as a penny dressed;
Sweet as a rose her breath and lips,
  Round as the globe her breast.

Full as an egg was I with glee,
  And happy as a king.
Good Lord! how all men envied me!
  She loved like anything.

But false as hell, she, like the wind,
  Changed, as her sex must do.
Though seeming as the turtle kind,
  And like the gospel true;

If I and Molly could agree
  Let who would take Peru!
Great as an Emperor should I be,
  And richer than a Jew;

Till you grow tender as a chick,
  I'm dull as any post;
Let us, like burrs, together stick,
  And warm as any toast.

You'll know me truer than a die,
  And wish me better sped;
Flat as a flounder when I lie,
  And as a herring dead.

Sure as a gun, she'll drop a tear
  And sigh perhaps, and wish,
When I am rotten as a pear,
  And mute as any fish.

ANONYMOUS

21

Drinking Song

SHE tells me with claret she cannot agree,
And she thinks of a hogshead whene'er she sees me;
For I smell like a beast, and therefore must I
Resolve to forsake her, or claret deny.
Must I leave my dear bottle, that was always my friend,
And I hope will continue so to my life's end?
Must I leave it for her? 'Tis a very hard task:
Let her go to the devil!—bring the other full flask.

Had she taxed me with gaming, and bid me forbear,
'Tis a thousand to one I had lent her an ear:
Had she found out my Sally, up three pair of stairs,
I had balked her, and gone to St. James's to prayers.
Had she bade me read homilies three times a day,
She perhaps had been humoured with little to say;
But, at night, to deny me my bottle of red,
Let her go to the devil!—there's no more to be said.

22

The Vicar of Bray

IN good King Charles's golden days,
When loyalty no harm meant,
A furious high-church man I was,
And so I gained preferment.
Unto my flock I daily preached,
Kings are by God appointed,
And damned are those who dare resist,
Or touch the Lord's anointed.
And this is law, I will maintain
Unto my dying day, Sir,
That whatsoever King shall reign,
I will be Vicar of Bray, Sir!

When Royal James possessed the crown,
And popery grew in fashion,
The penal law I hooted down,
And read the declaration:
The Church of Rome I found would fit
Full well my constitution,
And I had been a Jesuit,
But for the Revolution.
And this is law, etc.

ANONYMOUS

When William our deliverer came,
  To heal the nation's grievance,
I turned the cat in pan again,
  And swore to him allegiance:
Old principles I did revoke,
  Set conscience at a distance
Passive obedience is a joke,
  A jest is non-resistance.
    And this is law, etc.

When glorious Anne became our Queen,
  The Church of England's glory,
Another face of things was seen,
  And I became a Tory:
Occasional conformists base
  I damned, and moderation,
And thought the church in danger was
  From such prevarication.
    And this is law, etc.

When George in pudding time came o'er,
  And moderate men looked big, Sir,
My principles I changed once more,
  And so became a Whig, Sir:
And thus preferment I procured,
  From our Faith's Great Defender,
And almost every day abjured
  The Pope, and the Pretender.
    And this is law, etc.

The illustrious House of Hanover,
  And Protestant succession,
To these I lustily will swear,
  Whilst they can keep possession:
For in my faith and loyalty
  I never once will falter,
But George my lawful King shall be,
  Except the times should alter.
    And this is law, etc.

23

We're All Dry

WE'RE all dry with drinking on't,
We're all dry with drinking on't,
The piper kissed the fiddler's wife,
And I can't sleep for thinking on't.

JOHN BYROM

1692–1763

24

FROM *A full and true Account of A HORRID AND BARBAROUS ROBBERY, Committed in Epping Forest, on the Body of the Cambridge Coach*

DEAR Martin Folkes, dear scholar, brother, friend,
And words of like importance without end,
This comes to tell you, how in Epping Hundred
Last Wednesday morning I was robbed and plundered.
Forgive the Muse who sings what, I suppose,
Fame has already trumpeted in prose;
But Fame's a lying jade: the turn of fate
Let poor Melpomene herself relate:
Spare the sad nymph a vacant hour's relief,
To rhyme away the remnants of her grief.

On Tuesday night, you know with how much sorrow
I bid the Club farewell—'I go tomorrow'—
Tomorrow came, and so accordingly
Unto the place of rendezvous went I.
Bull was the house, and Bishopgate the street:
The coach as full as it could cram; to wit,
Two fellow commoners, *De Aula Trin*,
And eke an honest bricklayer of Lynn,
And eke two Norfolk dames, his wife and cousin,
And eke my worship's self made half a dozen.

Now then, as fortune had contrived, our way
Through the wild brakes of Epping Forest lay:
With travellers and trunks, a hugeous load,
We crawled along the solitary road;
Where naught but thickets within thickets grew,
No house nor barn to cheer the wandering view;
Nor labouring hind, nor shepherd did appear,
Nor sportsman with his dog or gun was there;
A dreary landscape, bushy and forlorn,
Where rogues start up like mushrooms in a morn.

However since we, none of us, had yet
Such rogues but in a sessions paper met,
We joked on fear; though as we passed along,
Robbing was still the burden of the song.
With untried courage bravely we repelled
The rude attacks of dogs—not yet beheld,
With valorous talk still battling, till at last
We thought all danger was as good as past.
Says one—too soon, alas—'Now let him come,
Full at his head I'll fling this bottle of rum.'

Scarce had he spoken when the brickman's wife
Cried out, 'Good Lord! he's here upon my life!'
Forth from behind the wheels the villain came,
And swore such words as I dare hardly name;
But you'll suppose them, brother, not to drop
From me, but him, 'G—d d—n ye, coachman, stop!
Your money, z—ds, deliver me your money,
Quick, d—n ye, quick; must I stay waiting on ye?
Quick, or I'll send'—(and nearer still he rode)
'A brace of balls amongst ye all, by G—d!'

I leave you, Sir, to judge yourself what plight
We all were put in by this cursed wight.
The trembling females into labour fell;
Big with the sudden fear, they pout, they swell;
And soon delivered by his horrid curses,
Brought forth two strange and preternatural purses;
That looked indeed like purses made of leather;
But let the sweet-tongued Maningham say whether
A common purse could possibly conceal
Shillings, half-crowns and ha'pence by piecemeal.

The youth who threw the bottle at the knave
Before he came, now thought it best to waive
Such resolution, and preserve the liquor;
Since a round guinea might be thrown much quicker:
So with impetuous haste he flung him that,
Which the sharp rascal parried with his hat.
His right-hand man, a brother of our quill,
Prudently chose to show his own good will
By the same token, and without much scruple
Made the red-rugged collector's income duple.

My heart—for truth I always must confess—
Dropped down an inch exactly more or less.
With both my eyes I viewed the thief's approach,
And read the case of Pistol versus Coach;
A woeful case which I had oft heard quoted,
But ne'er before in all my practice noted.
So when the lawyers brought in their report,
Guinea per Christian to be paid in court,
'Well off,' thinks I, 'from this son of a whore,
If he prefer his action for no more.'

'No more! why hang him; is not that much
To pay a guinea for his vile High-Dutch?
'Tis true his arguments are short and frank,
His action strong, to which he swears point-blank;
Yet why resign the yellow one pound one?
No, tax his bill and give him silver, John!
So said so done, when putting fist to fob
I flung the apparent value of the job,
An ounce of silver into his receiver,
And marked the issue of the rogue's behaviour.

He, like a thankless wretch that's overpaid,
Resents, forsooth, the affront upon his trade,
And treats my kindness with a 'This won't do;
Look ye here, Sir, I must ha' gold from you.'
To this demand of the ungrateful cur
Defendant John thought proper to demur.
The bricklayer, joining in the white opinion,
Tendered five shillings to Diana's minion,
Who still kept threatening to pervade his buff,
Because the payment was not prompt enough.

Before the women, with their purses each
Had strength to place contents within his reach,
One of his pieces falling downwards, drew
The rogue's attention hungrily thereto.
Straight he began to damn the charioteer:
'Come down, ye dog, reach me that guinea there.'
Down jumps the affrighted coachman on the sand,
Picks up the gold and puts it in his hand;
Missing a rare occasion, timorous dastard,
To seize his pistol and dismount the bastard.

Now while in deep and serious ponderment
I watched the motions of his next intent,
He wheeled about, as one full bent to try
The matter in dispute 'twixt him and I,
And how my silver sentiments would hold
Against that hard dilemma—balls or gold.
'No help?' said I, 'no tachygraphic power
To interpose in this unequal hour!
I doubt—I must resign—there's no defending
The cause against that murderous fire-engine.'

[*At this point the character of the poem changes, and the traveller is rescued by supernatural intervention*—Ed.]

25

Epigram

SOME say, compared to Bononcini,
That Mynheer Handel's but a ninny;
Others aver that he to Handel
Is scarcely fit to hold a candle.
Strange all this difference should be
'Twixt Tweedledum and Tweedledee!

26 *Extempore Intended to Allay the Violence of Party Spirit*

GOD bless the king, I mean the faith's defender;
God bless—(no harm in blessing)—the pretender;
But who pretender is, and who is king,
God bless us all—that's quite another thing.

27 *Four Epigrams on the Naturalization Bill*

(i)

COME all ye foreign strolling gentry,
Into Great Britain make your entry;
Abjure the Pope, and take the oaths,
And you shall have meat, drink, and clothes.

(ii)

WITH languages dispersed, men were not able
To top the skies, and build the tower of Babel;
But if to Britain they shall cross the main,
And meet by Act of Parliament again,
Who knows, when all together shall repair,
How high a castle may be built in air!

(iii)

THIS act reminds me, ge'men, under favour,
Of old John Bull, the hair-merchant and shaver:
John had a sign put up, whereof the writing
Was strictly copied from his own inditing:
Under the painted wigs both bob and full—
—*Moast munny pade for living* HERE—
John Bull.

(iv)

Now upon sale, a bankrupt island,
To any stranger that will buy land—
The birthright, note, for further satis-
Faction, is to be thrown in gratis.

BENJAMIN FRANKLIN

1706–1790

28

Impromptu

Jack, eating rotten cheese, did say,
Like Samson I my thousands slay:
I vow, quoth Roger, so you do,
And with the selfsame weapon too.

THOMAS LISLE

1709–1767

29

The Power of Music

When Orpheus went down to the regions below,
Which men are forbidden to see,
He tuned up his lyre, as old histories show,
To set his Eurydice free.

All hell was astonished a person so wise
Should rashly endanger his life,
And venture so far—but how vast their surprise
When they heard that he came for his wife.

To find out a punishment due to his fault
Old Pluto had puzzled his brain;
But hell had no torments sufficient, he thought,
So he gave him his wife back again.

But pity succeeding found place in his heart,
  And, pleased with his playing so well,
He took her again in reward of his art;
  Such power hath music in hell!

SAMUEL JOHNSON

1709–1784

(i)
Epigram

30

IF a man who turnips cries
Cry not when his father dies,
Is it not a proof he'd rather
Have a turnip than his father?

(ii)
Ballad

I PUT my hat upon my head
And walked into the Strand,
And there I met another man
Whose hat was in his hand.

(iii)
Idyll

HERMIT hoar, in solemn cell,
  Wearing out life's evening grey,
Strike thy bosom, Sage, and tell
  What is bliss, and which the way.

Thus I spoke, and speaking sighed,
  Scarce repressed the starting tear,
When the hoary sage replied,
  'Come, my lad, and drink some beer.'

ANONYMOUS

31 *The Sow Came in*

THE sow came in with the saddle,
The little pig rocked the cradle,
The dish jumped up on the table,
To see the pot swallow the ladle.
The spit that stood behind the door
Called the dish-clout dirty whore.
  Odd's-bobs! says the gridiron,
    Can't you agree?
  I'm the head constable,
    Bring them to me.

32 *Trip upon Trenchers*

TRIP upon trenchers, and dance upon dishes,
My mother sent me for some barm, some barm;
She bid me tread lightly, and come again quickly,
For fear the young men should do me some harm.
  Yet didn't you see, yet didn't you see,
  What naughty tricks they put upon me:
    They broke my pitcher,
    And spilt the water,
    And huffed my mother,
    And chid her daughter,
  And kissed my sister instead of me.

THOMAS GRAY

1716–1771

33

Ode on the Death of a Favourite Cat, Drowned in a Tub of Gold Fishes

'TWAS on a lofty vase's side,
Where China's gayest art had dyed
The azure flowers, that blow;
Demurest of the tabby kind,
The pensive Selima reclined,
Gazed on the lake below.

Her conscious tail her joy declared,
The fair round face, the snowy beard,
The velvet of her paws,
Her coat, that with the tortoise vies,
Her ears of jet and emerald eyes,
She saw, and purred applause.

Still had she gazed but 'midst the tide
Two angel forms were seen to glide,
The Genii of the stream:
Their scaly armour's Tyrian hue
Through richest purple to the view
Betrayed a golden gleam.

The hapless nymph with wonder saw:
A whisker first and then a claw,
With many an ardent wish,
She stretched in vain to reach the prize.
What female heart can gold despise?
What cat's averse to fish?

Presumptuous maid! with looks intent
Again she stretched, again she bent,
Nor knew the gulf between.
(Malignant Fate sat by, and smiled)
The slippery verge her feet beguiled,
She tumbled headlong in.

Eight times emerging from the flood
She mewed to every watry god,
  Some speedy aid to send.
No dolphin came, no Nereid stirred:
Nor cruel Tom, nor Susan heard.
  A favourite has no friend!

From hence, ye beauties, undeceived,
Know, one false step is ne'er retrieved,
  And be with caution bold.
Not all that tempts your wandering eyes
And heedless hearts, is lawful prize;
  Nor all that glisters, gold.

ANONYMOUS

34

On Prince Frederick

HERE lies Fred,
Who was alive and is dead:
Had it been his father,
I had much rather;
Had it been his brother,
Still better than another;
Had it been his sister,
No one would have missed her;
Had it been the whole generation,
So much the better for the nation:
But since 'tis only Fred,
Who was alive and is dead,
There's no more to be said.

35

Rub-a-dub-dub

Rub-a-dub-dub,
Three men in a tub,
And how do you think they got there?
The butcher, the baker,
The candlestick-maker,
They all jumped out of a rotten potato,
'Twas enough to make a man stare.

36

Tweedledum and Tweedledee

Tweedledum and Tweedledee
Agreed to have a battle,
For Tweedledum said Tweedledee
Had spoiled his nice new rattle.
Just then flew down a monstrous crow,
As black as a tar-barrel,
Which frightened both those heroes so,
They quite forgot their quarrel.

37

Baby, Baby

Baby, baby, naughty baby,
Hush, you squalling thing, I say.
Peace this moment, peace, or maybe
Bonaparte will pass this way.

Baby, baby, he's a giant,
Tall and black as Rouen steeple,
And he breakfasts, dines, rely on't,
Every day on naughty people.

Baby, baby, if he hear you,
As he gallops past the house,
Limb from limb at once he'll tear you,
Just as pussy tears a mouse.

And he'll beat you, beat you, beat you,
And he'll beat you all to pap,
And he'll eat you, eat you, eat you,
Every morsel snap, snap, snap.

ROBERT SOUTHEY

1774–1843

38

Ode to a Pig
while his Nose was being Bored

HARK! hark! that Pig—that Pig! the hideous note,
  More loud, more dissonant, each moment grows!
Would one not think the knife was in his throat?
  And yet they are only boring through his nose.

Pig! 'tis your master's pleasure—then be still,
  And hold your nose to let the iron through!
Dare you resist your lawful Sovereign's will?
  Rebellious Swine! you know not what you do.

To man o'er beast the sacred power was given;
  Pig, hear the truth, and never murmur more!
Would you rebel against the will of Heaven?
  You impious beast, be still, and let them bore!

The social Pig resigns his natural rights
  When first with man he covenants to live;
He barters them for safer sty delights,
  For grains and wash, which man alone can give.

Sure is provision on the social plan,
  Secure the comforts that to each belong!
Oh, happy Swine! the impartial sway of man
  Alike protects the weak Pig and the strong.

And you resist! you struggle now because
  Your master has thought fit to bore your nose!
You grunt in flat rebellion to the laws
  Society finds needful to impose!

Go to the forest, Piggy, and deplore
  The miserable lot of savage Swine!
See how the young Pigs fly from the great Boar,
  And see how coarse and scantily they dine!

Behold their hourly danger, when who will
  May hunt or snare or seize them for his food!
Oh, happy Pig! whom none presumes to kill
  Till your protecting master thinks it good!

And when, at last, the closing hour of life
  Arrives (for Pigs must die as well as Man),
When in your throat you feel the long sharp knife,
  And the blood trickles to the pudding-pan;

And when, at last, the death wound yawning wide,
  Fainter and fainter grows the expiring cry,
Is there no grateful joy, no loyal pride,
  To think that for your master's good you die?

39

To a Goose

IF thou didst feed on western plains of yore;
Or waddle wide with flat and flabby feet
Over some Cambrian mountain's plashy moor;
Or find in farmer's yard a safe retreat
From gypsy thieves, and foxes sly and fleet;
If thy grey quills, by lawyer guided, trace
Deeds big with ruin to some wretched race,
Or love-sick poet's sonnet, sad and sweet,
Wailing the rigour of his lady fair;
Or if, the drudge of housemaid's daily toil,
Cobwebs and dust thy pinions white besoil,
Departed Goose! I neither know nor care.
But this I know, that we pronounced thee fine,
Seasoned with sage and onions, and port wine.

EBENEZER ELLIOTT

1781–1849

40

On Communists

WHAT is a Communist? One who has yearnings
For equal division of unequal earnings;
Idler or bungler, or both, he is willing
To fork out his penny and pocket your shilling.

LEIGH HUNT

1784–1859

41

Three Sonnets

(i)

To a Fish

YOU strange, astonished-looking, angle-faced,
Dreary-mouthed, gaping wretches of the sea,
Gulping salt water everlastingly,
Cold-blooded, though with red your blood be graced,
And mute, though dwellers in the roaring waste;
And you, all shapes beside, that fishy be—
Some round, some flat, some long, all devilry,
Legless, unloving, infamously chaste:

O scaly, slippery, wet, swift, staring wights,
What is't ye do? What life lead? eh, dull goggles?
How do ye vary your vile days and nights?
How pass your Sundays? Are ye still but joggles
In ceaseless wash? Still nought but gapes, and bites,
And drinks, and stares, diversified with boggles?

(ii)

A Fish replies

Amazing monster! that, for aught I know,
With the first sight of thee didst make our race
For ever stare! O flat and shocking face,
Grimly divided from the breast below!
Thou that on dry land horribly dost go
With a split body and most ridiculous pace,
Prong after prong, disgracer of all grace,
Long-useless-finned, haired, upright, unwet, slow!

O breather of unbreathable, sword-sharp air,
How canst exist? How bear thyself, thou dry
And dreary sloth? What particle canst share
Of the only blessed life, the watery?
I sometimes see of ye an actual *pair*
Go by, linked fin by fin, most odiously.

(iii)

The Fish turns into a Man, and then into a Spirit, and again speaks

Indulge thy smiling scorn, if smiling still,
O man! and loathe, but with a sort of love;
For difference must its use by difference prove,
And, in sweet clang, the spheres with music fill.
One of the spirits am I, that at his will
Live in whate'er has life—fish, eagle, dove—
No hate, no pride, beneath naught, nor above,
A visitor of the rounds of God's sweet skill.

Man's life is warm, glad, sad, 'twixt loves and graves,
Boundless in hope, honoured with pangs austere,
Heaven-gazing; and his angel-wings he craves:
The fish is swift, small-needing, vague yet clear,
A cold, sweet, silver life, wrapped in round waves,
Quickened with touches of transporting fear.

THOMAS L. PEACOCK

1785–1866

42

Rich and Poor, or Saint and Sinner

THE poor man's sins are glaring;
In the face of ghostly warning
He is caught in the fact
Of an overt act—
Buying greens on Sunday morning.

The rich man's sins are hidden
In the pomp of wealth and station;
And escape the sight
Of the children of light,
Who are wise in their generation.

The rich man has a kitchen,
And cooks to dress his dinner;
The poor who would roast
To the baker's must post,
And thus becomes a sinner.

The rich man has a cellar,
And a ready butler by him;
The poor must steer
For his pint of beer
Where the saint can't choose but spy him.

The rich man's painted windows
Hide the concerts of the quality;
The poor can but share
A crack'd fiddle in the air,
Which offends all sound morality.

The rich man is invisible
In the crowd of his gay society;
But the poor man's delight
Is a sore in the sight,
And a stench in the nose of piety.

The rich man has a carriage
Where no rude eye can flout him;
    The poor man's bane
    Is a third-class train,
With the daylight all about him.

The rich man goes out yachting,
Where sanctity can't pursue him;
    The poor goes afloat
    In a fourpenny boat,
Where the bishop groans to view him.

ANONYMOUS

43 *The Night before Larry was Stretched*

THE night before Larry was stretched,
The boys they all paid him a visit;
A bait in their sacks too they fetched,
They sweated their duds till they riz it;
For Larry was always the lad,
When a friend was condemned to the squeezer,
Would fence all the togs that he had
Just to help the poor boy to a sneezer,
And moisten his gob 'fore he died.

The boys they came crowding in fast,
They drew all their stools round about him,
Six glims round his trap-case they placed;
He couldn't be well waked without 'em.
I axed was he fit for to die
Without having truly repented?
Says Larry, 'That's all in my eye,
And first by the clergy invented,
To get a fat bit for themselves.'

'I'm sorry now, Larry,' says I,
'To see you in this situation;
And, blister my limbs if I lie,
I'd as lief it had been my own station.'
'Och hone! 'tis all over,' says he,
'For the neck-cloth I'll be forced to put on,
And by this time tomorrow you'll see
Your Larry will be dead as mutton,
Because why, his courage was good.

'And I'll be cut up like a pie,
And my knob from my body be parted.'
'You're in the wrong box then,' says I,
'For blast me if they're so hard-hearted;
A chalk on the back of your neck
Is all that Jack Ketch dares to give you;
Then mind not such trifles a feck,
For why should the likes of them grieve you?
And now boys, come tip us the deck.'

The cards being called for, they played,
Till Larry found one of them cheated;
A dart at his napper he made
(The boy being easily heated);
'Oh! by the hokey, you thief,
I'll scuttle your knob with my daddle!
You cheat me because I'm in grief,
But soon I'll demolish your noddle,
And leave you your claret to drink.'

Then the clergy came in with his book,
He spoke him so smooth and so civil;
Larry tipped him a Kilmainham look,
And pitched his big wig to the devil;
Then stooping a little his head,
To get a sweet drop of the bottle,
And pitiful sighing, he said:
'Oh, the hemp will be soon round my throttle,
And choke my poor windpipe to death.

'Though sure it's the best way to die,
Oh! the devil a better a-livin'!
For when the gallows is high,
Your journey is shorter to heaven.'
But what harasses Larry the most,
And makes his poor soul melancholy,
Is he thinks of the time when his ghost
Will come in a sheet to sweet Molly;
'Oh, sure it will kill her alive!'

So moving these last words he spoke,
We all vented our tears in a shower;
For my part I thought my heart broke,
To see him cut down like a flower.
On his travels we watched him next day;
Oh, the throttler, I thought I could kill him;
But Larry not one word did say,
Nor changed till he came to King William,
Then musha, his colour turned white.

When he came to the nubbling chit,
He was tucked up so neat and so pretty;
The rumbler jogged off from his feet,
And he died with his face to the city;
He kicked too, but that was all pride,
For soon you might see 'twas all over;
Soon after, the noose was untied,
And at darkee we waked him in clover,
And sent him to take a ground sweat.

44

In Peterborough Churchyard

READER, pass on, nor idly waste your time
In bad biography, or bitter rhyme;
For what I am, this cumbrous clay ensures,
And what I was is no affair of yours.

GEORGE GORDON NOEL, LORD BYRON

1788–1824

45 FROM *Beppo*

'Tis known, at least it should be, that throughout
  All countries of the Catholic persuasion,
Some weeks before Shrove Tuesday comes about,
  The people take their fill of recreation,
And buy repentance, ere they grow devout,
  However high their rank, or low their station,
With fiddling, feasting, dancing, drinking, masquing,
And other things which may be had for asking.

The moment night with dusky mantle covers
  The skies (and the more duskily the better),
The time less liked by husbands than by lovers
  Begins, and prudery flings aside her fetter;
And gaiety on restless tiptoe hovers,
  Giggling with all the gallants who beset her;
And there are songs and quavers, roaring, humming,
Guitars, and every other sort of strumming.

And there are dresses splendid, but fantastical,
  Masks of all times and nations, Turks and Jews,
And harlequins and clowns, with feats gymnastical,
  Greeks, Romans, Yankee-doodles, and Hindoos;
All kinds of dress, except the ecclesiastical,
  All people, as their fancies hit, may choose,
But no one in these parts may quiz the clergy—
Therefore take heed, ye Freethinkers! I charge ye.

You'd better walk about begirt with briars,
  Instead of coat and smallclothes, than put on
A single stitch reflecting upon friars,
  Although you swore it only was in fun;
They'd haul you o'er the coals, and stir the fires
  Of Phlegethon with every mother's son,
Nor say one mass to cool the cauldron's bubble
That boil'd your bones, unless you paid them double.

But saving this, you may put on whate'er
  You like by way of doublet, cape, or cloak,
Such as in Monmouth-street, or in Rag Fair,
  Would rig you out in seriousness or joke;
And even in Italy such places are,
  With prettier name in softer accents spoke,
For, bating Covent Garden, I can hit on
No place that's called 'Piazza' in Great Britain.

This feast is named the Carnival, which being
  Interpreted, implies 'farewell to flesh':
So call'd, because the name and thing agreeing,
  Through Lent they live on fish both salt and fresh.
But why they usher Lent with so much glee in,
  Is more than I can tell, although I guess
'Tis as we take a glass with friends at parting,
In the stage-coach or packet, just at starting.

And thus they bid farewell to carnal dishes,
  And solid meats, and highly spiced ragouts,
To live for forty days on ill-dress'd fishes,
  Because they have no sauces to their stews;
A thing which causes many 'poohs' and 'pishes,'
  And several oaths (which would not suit the Muse),
From travellers accustom'd from a boy
To eat their salmon, at the least, with soy;

And therefore humbly I would recommend
  'The curious in fish-sauce,' before they cross
The sea, to bid their cook, or wife, or friend,
  Walk or ride to the Strand, and buy in gross
(Or if set out beforehand, these may send
  By any means least liable to loss),
Ketchup, Soy, Chili-vinegar, and Harvey,
Or, by the Lord! a Lent will well nigh starve ye;

That is to say, if your religion's Roman,
  And you at Rome would do as Romans do,
According to the proverb although no man,
  If foreign, is obliged to fast; and you
If Protestant, or sickly, or a woman,
  Would rather dine in sin on a ragout—
Dine and be d—d! I don't mean to be coarse,
But that's the penalty, to say no worse.

GEORGE GORDON NOEL, LORD BYRON

Of all the places where the Carnival
  Was most facetious in the days of yore,
For dance, and song, and serenade, and ball,
  And masque, and mime, and mystery, and more
Than I have time to tell now, or at all,
  Venice the bell from every city bore—
And at the moment when I fix my story,
That sea-born city was in all her glory.

They've pretty faces yet, those same Venetians,
  Black eyes, arch'd brows, and sweet expressions
      still;
Such as of old were copied from the Grecians,
  In ancient arts by moderns mimick'd ill;
And like so many Venuses of Titian's
  (The best's at Florence—see it, if ye will,)
They look when leaning over the balcony,
Or stepp'd from out a picture by Giorgione,

Whose tints are truth and beauty at their best;
  And when you to Manfrini's palace go,
That picture (howsoever fine the rest)
  Is loveliest to my mind of all the show;
It may perhaps be also to *your* zest
  And that's the cause I rhyme upon it so:
'Tis but a portrait of his son, and wife,
And self; but *such* a woman! love in life!

Love in full life and length, not love ideal,
  No, nor ideal beauty, that fine name.
But something better still, so very real,
  That the sweet model must have been the same;
A thing that you would purchase, beg, or steal,
  Were't not impossible, besides a shame:
The face recalls some face, as 'twere with pain,
You once have seen, but ne'er will see again;

One of those forms which flit by us, when we
  Are young, and fix our eyes on every face;
And, oh! the loveliness at times we see
  In momentary gliding, the soft grace,

The youth, the bloom, the beauty which agree,
  In many a nameless being we retrace,
Whose course and home we knew not, nor shall know,
Like the lost Pleiad seen no more below.

I said that like a picture by Giorgione
  Venetian women were, and so they *are*,
Particularly seen from a balcony,
  (For beauty's sometimes best set off afar)
And there, just like a heroine of Goldoni,
  They peep from out the blind, or o'er the bar;
And truth to say, they're mostly very pretty,
And rather like to show it, more's the pity!

For glances beget ogles, ogles sighs,
  Sighs wishes, wishes words, and words a letter,
Which flies on wings of light-heel'd Mercuries,
  Who do such things because they know no better;
And then, God knows, what mischief may arise,
  When love links two young people in one fetter,
Vile assignations, and adulterous beds,
Elopements, broken vows, and hearts, and heads.

Shakespeare described the sex in Desdemona
  As very fair, but yet suspect in fame,
And to this day from Venice to Verona
  Such matters may be probably the same,
Except that since those times was never known a
  Husband whom mere suspicion could inflame
To suffocate a wife no more than twenty,
Because she had a 'cavalier servente.'

Their jealousy (if they are ever jealous)
  Is of a fair complexion altogether,
Not like that sooty devil of Othello's
  Which smothers women in a bed of feather,
But worthier of these much more jolly fellows,
  When weary of the matrimonial tether
His head for such a wife no mortal bothers,
But takes at once another or another's.

Didst ever see a Gondola? For fear
  You should not, I'll describe it you exactly:
'Tis a long cover'd boat that's common here,
  Carved at the prow, built lightly, but compactly,
Row'd by two rowers, each call'd 'Gondolier,'
  It glides along the water looking blackly,
Just like a coffin clapt in a canoe,
Where none can make out what you say or do.

And up and down the long canals they go,
  And under the Rialto shoot along,
By night and day, all paces, swift or slow,
  And round the theatres, a sable throng,
They wait in their dusk livery of woe—
  But not to them do woeful things belong,
For sometimes they contain a deal of fun,
Like mourning coaches when the funeral's done.

46 FROM *Don Juan*

SAGEST of women, even of widows, she
Resolved that Juan should be quite a paragon,
And worthy of the noblest pedigree:
  (His sire was of Castile, his dam from Aragon).
Then for accomplishments of chivalry,
  In case our lord the king should go to war again,
He learned the arts of riding, fencing, gunnery,
And how to scale a fortress—or a nunnery.

But that which Donna Inez most desired,
  And saw into herself each day before all
The learned tutors whom for him she hired,
  Was, that his breeding should be strictly moral:
Much into all his studies she inquired,
  And so they were submitted first to her, all
Arts, sciences, no branch was made a mystery
To Juan's eyes, excepting natural history.

The languages, especially the dead,
  The sciences, and most of all the abstruse,
The arts, at least all such as could be said
  To be the most remote from common use,
In all these he was much and deeply read:
  But not a page of anything that's loose
Or hints continuation of the species,
Was ever suffered, lest he should grow vicious.

His classic studies made a little puzzle,
  Because of filthy loves of gods and goddesses,
Who in the earlier ages raised a bustle,
  But never put on pantaloons or bodices;
His reverend tutors had at times a tussle,
  And for their Aeneids, Iliads, and Odysseys
Were forced to make an odd sort of apology,
For Donna Inez dreaded the mythology.

Ovid's a rake, as half his verses show him,
  Anacreon's morals are a still worse sample,
Catullus scarcely has a decent poem,
  I don't think Sappho's ode a good example,
Although Longinus tells us there is no hymn
  Where the sublime soars forth on wings more ample;
But Virgil's songs are pure, except that horrid one
Beginning with 'Formosum pastor Corydon'.

Lucretius' irreligion is too strong
  For early stomachs, to prove wholesome food.
I can't help thinking Juvenal was wrong,
  Although no doubt his real intent was good,
For speaking out so plainly in his song,
  So much indeed as to be downright rude;
And then what proper person can be partial
To all those nauseous epigrams of Martial?

Juan was taught from out the best edition,
  Expurgated by learnèd men, who place
Judiciously from out the schoolboy's vision
  The grosser parts, but fearful to deface
Too much their modest bard by this omission,
  And pitying sore his mutilated case,
They only add them all in an appendix,
Which saves in fact the trouble of an index;

For there we have them all at one fell swoop,
  Instead of being scattered through the pages;
They stand forth marshalled in a handsome troop,
  To meet the ingenuous youth of future ages,
Till some less rigid editor shall stoop
  To call them back into their separate cages,
Instead of standing staring altogether,
Like garden gods—and not so decent either.

*

Milton's the prince of poets—so we say;
  A little heavy, but no less divine:
An independent being in his day—
  Learnèd, pious, temperate in love and wine;
But his life falling into Johnson's way,
  We're told this great high priest of all the Nine
Was whipt at college—a harsh sire—odd spouse,
For the first Mrs. Milton left his house.

All these are, *certes*, entertaining facts,
  Like Shakspeare's stealing deer, Lord Bacon's bribes;
Like Titus' youth and Caesar's earliest acts;
  Like Burns (whom Doctor Currie well describes);
Like Cromwell's pranks; but although truth exacts
  These amiable descriptions from the scribes,
As most essential to their hero's story,
They do not much contribute to his glory.

All are not moralists, like Southey, when
  He prated to the world of 'Pantisocracy';
Or Wordsworth unexcised, unhired, who then
  Seasoned his pedlar poems with democracy;
Or Coleridge, long before his flighty pen
  Let to the Morning Post its aristocracy;
When he and Southey, following the same path,
Espoused two partners (milliners of Bath).

Such names at present cut a convict figure,
  The very Botany Bay in moral geography;
Their loyal treason, renegado rigour,
  Are good manure for their more bare biography,
Wordsworth's last quarto, by the way, is bigger
  Than any since the birthday of typography;
A drowsy frowzy poem, called the 'Excursion,'
Writ in a manner which is my aversion.

He there builds up a formidable dyke
  Between his own and others' intellect:
But Wordsworth's poem and his followers, like
  Joanna Southcote's Shiloh and her sect,
Are things which in this century don't strike
  The public mind—so few are the elect;
And the new births of both their stale virginities
Have proved but dropsies, taken for divinities.

But let me to my story: I must own,
  If I have any fault, it is digression,
Leaving my people to proceed alone,
  While I soliloquize beyond expression:
But these are my addresses from the throne,
  Which put off business to the ensuing session:
Forgetting each omission is a loss to
The world, not quite so great as Ariosto.

I know that what our neighbours call *longueurs*,
  (We've not so good a word, but have the thing,
In that complete perfection which ensures
  An epic from Bob Southey every Spring—)
Form not the true temptation which allures
  The reader; but 'twould not be hard to bring
Some fine examples of the *épopée*,
To prove its grand ingredient is *ennui*.

We learn from Horace, 'Homer sometimes sleeps';
  We feel without him Wordsworth sometimes wakes—
To show with what complacency he creeps,
  With his dear '*Waggoners*', around his lakes.
He wishes for 'a boat' to sail the deeps.
  Of ocean?—No, of air. And then he makes
Another outcry for 'a little boat',
And drivels seas to set it well afloat.

If he must fain sweep o'er the ethereal plain,
  And Pegasus runs restive in his 'Waggon',
Could he not beg the loan of Charles's Wain?
  Or pray Medea for a single dragon?
Or if, too classic for his vulgar brain,
  He feared his neck to venture such a nag on,
And he must needs mount nearer to the moon,
Could not the blockhead ask for a balloon?

'Pedlars', and 'boats', and 'waggons'! Oh! ye shades
  Of Pope and Dryden, are we come to this?
That trash of such sort not alone evades
  Contempt, but from the bathos' vast abyss
Floats scumlike uppermost, and these Jack Cades
  Of sense and song above your graves may hiss—
The 'little boatman' and his 'Peter Bell'
Can sneer at him who drew 'Achitophel'!

*

Over the stones still rattling, up Pall Mall,
  Through crowds and carriages, but waxing thinner
As thunder'd knockers broke the long-seal'd spell
  Of doors 'gainst duns, and to an early dinner
Admitted a small party as night fell—
  Don Juan, our young diplomatic sinner,
Pursued his path, and drove past some hotels,
St. James's Palace and St. James's 'hells'.

They reach'd the hotel: forth streamed from the front
    door
  A tide of well-clad waiters, and around
The mob stood and as usual several score
  Of those pedestrian Paphians who abound
In decent London when the daylight's o'er.
  Commodious but immoral, they are found
Useful, like Malthus, in promoting marriage—
But Juan now is stepping from his carriage

Into one of the sweetest of hotels,
  Especially for foreigners—and mostly
For those whom favour or whom fortune swells,
  And cannot find a bill's small items costly.
There many an envoy either dwelt or dwells
  (The den of many a diplomatic lost lie),
Until to some conspicuous square they pass
And blazon o'er the door their names in brass.

Juan, whose was a delicate commission,
  Private though publicly important, bore
No title to point out with due precision
  The exact affair on which he was sent o'er.
'Twas merely known that on a secret mission
  A foreigner of rank had graced our shore,
Young, handsome, and accomplish'd, who was said
(In whispers) to have turned his sovereign's head.

Some rumour also of some strange adventures
  Had gone before him, and his wars and loves;
And as romantic heads are pretty painters,
  And, above all, an Englishwoman's roves
Into the excursive, breaking the indentures
  Of sober reason, wheresoe'er it moves,
He found himself extremely in the fashion,
Which serves our thinking people for a passion.

I don't mean that they are passionless, but quite
  The contrary, but then 'tis in the head;
Yet as the consequences are as bright
  As if they acted with the heart instead,
What after all can signify the site
  Of ladies' lucubrations? So they lead
In safety to the place for which you start,
What matters if the road be head or heart?

Juan presented in the proper place,
  To proper placemen, every Russ credential,
And was received with all the due grimace
  By those who govern in the mood potential,
Who, seeing a handsome stripling with smooth face,
  Thought (what in state affairs is most essential)
That they as easily might *do* the youngster,
As hawks may pounce upon a woodland songster.

They err'd, as aged men will do, but by
  And by we'll talk of that, and if we don't,
'Twill be because our notion is not high
  Of politicians and their double front,
Who live by lies, yet dare not boldly lie:—
  Now what I love in women is, they won't
Or can't do otherwise than lie, but do it
So well, the very truth seems falsehood to it.

And, after all what is a lie? 'Tis but
  The truth in masquerade, and I defy
Historians, heroes, lawyers, priests, to put
  A fact without some leaven of a lie.
The very shadow of true truth would shut
  Up annals, revelations, poesy,
And prophecy—except it should be dated
Some years before the incidents related.

Praised be all liars and all lies! Who now
  Can tax my mild Muse with misanthropy?
She rings the world's 'Te Deum', and her brow
  Blushes for those who will not:—but to sigh
Is idle. Let us like most others bow,
  Kiss hands, feet, any part of majesty,
After the good example of 'Green Erin',
Whose shamrock now seems rather worse for wearing.

Don Juan was presented, and his dress
  And mien excited general admiration—
I don't know which was most admired or less:
  One monstrous diamond drew much observation,
Which Catherine in a moment of '*ivresse*'
  (In love or brandy's fervent fermentation)
Bestow'd upon him, as the public learn'd;
And, to say truth, it had been fairly earn'd.

Besides the ministers and underlings,
  Who must be courteous to the accredited
Diplomatists of rather wavering kings,
  Until their royal riddle's fully read,
The very clerks—those somewhat dirty springs
  Of office, or the house of office, fed
By foul corruption into streams—even they
Were hardly rude enough to earn their pay.

And insolence no doubt is what they are
  Employ'd for, since it is their daily labour,
In the dear offices of peace or war;
  And should you doubt, pray ask of your next neighbour,
When for a passport, or some other bar
  To freedom, he applied (a grief and a bore),
If he found not in this spawn of tax born riches,
Like lap dogs, the least civil sons of bitches.

*

A young unmarried man, with a good name
  And fortune, has an awkward part to play;
For good society is but a game,
  'The royal game of Goose', as I may say,
Where everybody has some separate aim,
  An end to answer, or a plan to lay—
The single ladies wishing to be double,
The married ones to save the virgins trouble.

I don't mean this as general, but particular
  Examples may be found of such pursuits:
Though several also keep their perpendicular
  Like poplars, with good principles for roots;
Yet many have a method more *reticular*—
  'Fishers for men', like sirens with soft lutes:
For talk six times with the same single lady,
And you may get the wedding dresses ready.

Perhaps you'll have a letter from the mother,
  To say her daughter's feelings are trepann'd;
Perhaps you'll have a visit from the brother,
  All strut, and stays, and whiskers, to demand
What 'your intentions are?'—One way or other
  It seems the virgin's heart expects your hand:
And between pity for her case and yours,
You'll add to Matrimony's list of cures.

I've known a dozen weddings made even *thus*,
  And some of them high names: I have also known
Young men who—though they hated to discuss
  Pretensions which they never dreamed to have shown—
Yet neither frighten'd by a female fuss,
  Nor by mustachios moved, were let alone,
And lived, as did the brokenhearted fair,
In happier plight than if they form'd a pair.

There's also nightly, to the uninitiated,
  A peril—not indeed like love or marriage,
But not the less for this to be depreciated:
  It is—I meant and mean not to disparage
The show of virtue even in the vitiated—
  It adds an outward grace unto their carriage—
But to denounce the amphibious sort of harlot,
Couleur de rose, who's neither white nor scarlet.

Such is your cold coquette, who can't say 'no',
  And won't say 'yes', and keeps you on- and off-ing
On a lee shore, till it begins to blow—
  Then sees your heart wreck'd with an inward scoffing.
This works a world of sentimental woe,
  And sends new Werters yearly to their coffin;
But yet is merely innocent flirtation,
Not quite adultery, but adulteration.

*

You know, or don't know, that great Bacon saith,
  'Fling up a straw, 'twill show the way the wind blows';
And such a straw, borne on by human breath,
  Is poesy, according as the mind glows;
A paper kite, which flies 'twixt life and death,
  A shadow which the onward soul behind throws:
And mine's a bubble, not blown up for praise,
But just to play with, as an infant plays.

The world is all before me—or behind;
  For I have seen a portion of that same,
And quite enough for me to keep in mind—
  Of passions, too, I have proved enough to blame,
To the great pleasure of our friends, mankind,
  Who like to mix some slight alloy with fame;
For I was rather famous in my time,
Until I fairly knock'd it up with rhyme.

I have brought this world about my ears, and eke
  The other; that's to say, the clergy—who
Upon my head have bid their thunders break
  In pious libels by no means a few.
And yet I can't help scribbling once a week,
  Tiring old readers, nor discovering new.
In youth I wrote because my mind was full,
And *now* because I feel it growing dull.

But 'why then publish?'—There are no rewards,
  Of fame or profit when the world grows weary.
I ask in turn—why do you play at cards?
  Why drink? Why read?—To make some hour less
    dreary.
It occupies me to turn back regards
  On what I've seen or ponder'd, sad or cheery;
And what I write I cast upon the stream,
To swim or sink—I have had at least my dream.

I think that were I *certain* of success,
  I hardly could compose another line:
So long I've battled either more or less,
  That no defeat can drive me from the Nine.
This feeling 'tis not easy to express,
  And yet 'tis not affected, I opine.
In play there are two pleasures for your choosing—
The one is winning, and the other losing.

Besides, my Muse by no means deals in fiction:
  She gathers a repertory of facts,
Of course with some reserve and slight restriction,
  But mostly sings of human things and acts—
And that's one cause she meets with contradiction;
  For too much truth, at first sight ne'er attracts;
And were her object only what's called glory,
With more ease too she'd tell a different story.

Love, war, a tempest—surely there's variety:
  Also a seasoning slight of lucubration;
A bird's-eye view, too, of that wild, Society;
  A slight glance thrown on men of every station.
If you have naught else, here's at least satiety,
  Both in performance and in preparation;
And though these lines should only line portmanteaus,
Trade will be all the better for these Cantos.

The portion of this world which I at present
  Have taken up to fill the following sermon,
Is one of which there's no description recent:
  The reason why is easy to determine:
Although it seems both prominent and pleasant,
  There is a sameness in its gems and ermine,
A dull and family likeness through all ages,
Of no great promise for poetic pages.

With much to excite, there's little to exalt;
  Nothing that speaks to all men and all times;
A sort of varnish over every fault;
  A kind of common-place even in their crimes,
Factitious passions, wit without much salt,
  A want of that true nature which sublimes
Whate'er it shows with truth; a smooth monotony
Of character, in those at least who have got any.

Sometimes, indeed, like soldiers off parade,
  They break their ranks and gladly leave the drill;
But then the roll-call draws them back afraid,
  And they must be or seem what they were: still
Doubtless it is a brilliant masquerade:
  But when of the first sight you have had your fill,
It palls—at least it did so upon me,
This paradise of pleasure and *ennui*.

47

Stanza: A Fragment

I WOULD to heaven that I were so much clay
  As I am blood, bone, marrow, passion, feeling—
For then at least the past were pass'd away—
  And for the future—(but I write this reeling,
Having got drunk exceedingly today,
  So that I seem to stand upon the ceiling)
I say—the future is a serious matter—
And so—for God's sake—hock and soda-water!

HARTLEY COLERIDGE

1796–1849

48

Wordsworth Unvisited

HE lived amidst th' untrodden ways
  To Rydal Lake that lead;—
A bard whom there were none to praise,
  And very few to read.

Behind a cloud his mystic sense,
  Deep-hidden, who can spy?
Bright as the night when not a star
  Is shining in the sky.

Unread his works—his 'Milk White Doe'
  With dust is dark and dim;
It's still in Longman's shop, and oh!
  The difference to him!

JAMES PLANCHÉ

1796–1880

49 *Love, You've been a Villain*

LOVERS who are young indeed, and wish to know the sort of life
That in this world you're like to lead, ere you can say you've caught a wife,
Listen to the lay of one who's had with Cupid much to do,
And love-sick once, is love-sick still, but in another point of view.
Woman, though so kind she seems, will take your heart and tantalize it,
Were it made of Portland stone, she'd manage to McAdamize it.
Dairy-maid or duchess,
Keep it from her clutches,
If you'd ever wish to know a quiet moment more.
Wooing, cooing,
Seeming, scheming,
Smiling, wiling,
Pleasing, teasing,
Taking, breaking,
Clutching, touching
Bosoms to the core.
Oh love you've been a villain since the days of Troy and Helen,
When you caused the fall of Paris and of very many more.
Sighing like a furnace, in the hope that you may win her still,
And losing health and appetite, and growing thin and thinner still;
Walking in the wet before her window or her door o'nights,
And catching nothing but a cold with waiting there a score o'nights.
Spoiling paper by the ream with rhymes devoid of reasoning,
As silly and insipid as a goose without the seasoning.
Running bills with tailors,
Locking up by jailers,
Bread and water diet then, your senses to restore.
Sighing, crying,
Losing, musing,
Walking, stalking,
Hatching, catching,
Spoiling, toiling,
Rhyming, chiming,
Running up a score.

Finding all you've suffered has but been the sport of jilting blades,
And calling out your rival in the style of all true tilting blades;
Feeling, ere you've breakfasted, a bullet through your body pass,
And cursing then your cruel fate, and looking very like an ass;
Popped into a coffin just as dead as suits your time of life;
Paragraphed in papers too, as 'cut off in the prime of life.'
When the earth you're under,
Just a nine days' wonder,
And the world jogs on again exactly as before.
Jilting, tilting,
Calling, falling,
Swearing, tearing,
Lying, dying,
Cenotaph'd, and paragraph'd,
And reckon'd quite a bore.

50 *Ching a Ring*

CHING a ring, a ring ching,
Feast of Lantherns.
What a crop of chop-sticks, hongs and gongs,
Hundred thousand Chinese
Crinkums crankums
Hung among the bells and ding dongs!
What a lot of Pekin pots and pipkins,
Mandarins with pigtails, rings and strings,
Funny little slop-shops, places, cases,
All among the cups and tea-things!
Women with their ten toes tight tuck'd into
Tiddle toddle shoes one scarcely sees;
How they all come there's quite a wonder,—
China must be broken in pieces.
Ching a ring, a ring ching,
Feast of Lantherns,
What a crop of chop-sticks, hongs and gongs,
Hundred thousand Chinese
Crinkums crankums,
Pekin pigtails, slop-shops, tea-things,
Hong gongs, ding dongs.
Pots and pipkins!—
China must be broken in pieces.

ALARIC A. WATTS

1797–1864

51

An Austrian Army

AN Austrian army awfully array'd,
Boldly by battery besieged Belgrade.
Cossack commanders cannonading come
Dealing destruction's devastating doom:
Every endeavour engineers essay,
For fame, for fortune fighting-furious fray!
Generals 'gainst generals grapple, gracious God!
How Heaven honours heroic hardihood!
Infuriate—indiscriminate in ill—
Kinsmen kill kindred—kindred kinsmen kill:
Labour low levels loftiest, longest lines,
Men march 'mid mounds, 'mid moles, 'mid murd'rous
  mines:
Now noisy noxious numbers notice nought
Of outward obstacles, opposing ought—
Poor patriots—partly purchased—partly press'd
Quite quaking, quickly 'Quarter! quarter!' quest:
Reasons returns, religious right redounds,
Suwarrow stops such sanguinary sounds.
Truce to thee, Turkey, triumph to thy train,
Unwise, unjust, unmerciful Ukraine!
Vanish, vain victory! Vanish, victory vain!
Why wish we warfare? Wherefore welcome were
Xerxes, Ximenes, Xanthus, Xavier?
Yield, yield, ye youths, ye yeomen, yield your yell:
Zeno's, Zimmermann's, Zoroaster's zeal,
Again attract; arts against arms appeal!

THOMAS HOOD

1799–1845

52 *Faithless Sally Brown*

YOUNG Ben he was a nice young man,
  A carpenter by trade;
And he fell in love with Sally Brown,
  That was a lady's maid.

But as they fetch'd a walk one day,
  They met a press-gang crew;
And Sally she did faint away,
  Whilst Ben he was brought to.

The Boatswain swore with wicked words,
  Enough to shock a saint,
That though she did seem in a fit,
  'Twas nothing but a feint.

'Come, girl,' said he, 'hold up your head,
  He'll be as good as me;
For when your swain is in our boat,
  A boatswain he will be.'

So when they'd made their game of her,
  And taken off her elf,
She roused, and found she only was
  A coming to herself.

'And is he gone, and is he gone?'
  She cried, and wept outright:
'Then I will to the water side,
  And see him out of sight.'

A waterman came up to her,
  'Now, young woman,' said he,
'If you weep on so, you will make
  Eye-water in the sea.'

'Alas! they've taken my beau Ben,
To sail with old Benbow;'
And her woe began to run afresh,
As if she'd said, Gee woe!

Says he, 'They've only taken him
To the Tender-ship, you see;'
'The Tender-ship,' cried Sally Brown,
'What a hard-ship that must be!

'Oh! would I were a mermaid now,
For then I'd follow him;
But Oh!—I'm not a fish-woman,
And so I cannot swim.

'Alas! I was not born beneath
The virgin and the scales,
So I must curse my cruel stars,
And walk about in Wales.'

Now Ben had sail'd to many a place
That's underneath the world;
But in two years the ship came home,
And all her sails were furl'd.

But when he call'd on Sally Brown,
To see how she got on,
He found she'd got another Ben,
Whose Christian-name was John.

'Oh, Sally Brown, oh, Sally Brown,
How could you serve me so,
I've met with many a breeze before,
But never such a blow!'

Then reading on his 'bacco box,
He heaved a heavy sigh,
And then began to eye his pipe,
And then to pipe his eye.

And then he tried to sing 'All's Well,'
But could not, though he tried;
His head was turn'd, and so he chew'd
His pigtail till he died.

His death, which happen'd in his berth,
  At forty-odd befell:
They went and told the sexton, and
  The sexton toll'd the bell.

53 *Faithless Nelly Gray*

BEN BATTLE was a soldier bold,
  And used to war's alarms;
But a cannon-ball took off his legs,
  So he laid down his arms!

Now as they bore him off the field,
  Said he, 'Let others shoot,
For here I leave my second leg,
  And the Forty-second Foot!'

The army-surgeons made him limbs:
  Said he—'They're only pegs:
But there's as wooden members quite
  As represent my legs!'

Now Ben he loved a pretty maid,
  Her name was Nelly Gray;
So he went to pay her his devours
  When he'd devoured his pay!

But when he called on Nelly Gray,
  She made him quite a scoff;
And when she saw his wooden legs
  Began to take them off!

'Oh, Nelly Gray! Oh, Nelly Gray!
  Is this your love so warm?
The love that loves a scarlet coat
  Should be more uniform!'

Said she, 'I loved a soldier once,
  For he was blithe and brave;
But I will never have a man
  With both legs in the grave!

'Before you had those timber toes,
  Your love I did allow,
But then, you know, you stand upon
  Another footing now!'

'Oh, Nelly Gray! Oh, Nelly Gray!
  For all your jeering speeches,
At duty's call, I left my legs
  In Badajos's *breaches*!'

'Why then,' said she, 'you've lost the feet
  Of legs in war's alarms,
And now you cannot wear your shoes
  Upon your feats of arms!'

'Oh, false and fickle Nelly Gray;
  I know why you refuse:—
Though I've no feet—some other man
  Is standing in my shoes!

'I wish I ne'er had seen your face;
  But, now, a long farewell!
For you will be my death—alas!
  You will not be my *Nell*!'

Now when he went from Nelly Gray,
  His heart so heavy got—
And life was such a burthen grown,
  It made him take a knot!

So round his melancholy neck,
  A rope he did entwine,
And, for his second time in life,
  Enlisted in the Line!

One end he tied around a beam,
  And then removed his pegs,
And, as his legs were off,—of course,
  He soon was off his legs!

And there he hung, till he was dead
  As any nail in town,—
For though distress had cut him up,
  It could not cut him down!

A dozen men sat on his corpse,
  To find out why he died—
And they buried Ben in four cross-roads,
  With a *stake* in his inside!

54

To Minerva

My temples throb, my pulses boil,
  I'm sick of Song, and Ode, and Ballad—
So, Thyrsis, take the Midnight Oil,
  And pour it on a lobster salad.

My brain is dull, my sight is foul,
  I cannot write a verse, or read,—
Then, Pallas, take away thine Owl,
  And let us have a lark instead.

ANONYMOUS

55

Mr. and Mrs. Vite's Journey

A vorthy cit, von Vitsunday,
Vith vife, rode out in vone-horse chay;
And down the streets as they did trot,
Says Mrs. Vite, 'I'll tell you vat,
    Dear Villiam Vite,
    'Tis my delight,
Ven our veek's bills ve stick 'em,
    That side by side
    Ve thus should ride
To Vindsor or Vest Vickham.'

'My loving vife, full vell you know,
Ve used to ride to Valthamstow,
And now I think it much the best
That ve should ride tovards the vest.
  If you agree,
  Dear vife, vith me,
And vish to change the scene;
  And, ven the dust
  Excites our thirst,
Ve'll stop at Valham Green.'

'Oh, then,' says Mrs. Vite, says she,
'Vat pleases you, must sure please me;
But veekly vorkins all must go,
If ve this day go cheerful through:
  For vel I loves
  The voods and groves,
They raptures put me in;
  For you know, Vite,
  Von Vitsun night,
You did my poor heart vin.'

Then Mrs. Vite she took the vip,
And vack'd poor Dobbin on the hip;
Vich made him from a valk run fast,
And reach the long-vish't sign at last.
  Lo, ven they stopt,
  Out vaiter popt,
'Vat vou'd you vish to take?'
  Said Vite, vith grin,
  'I'll take some gin,
My vife takes vine and cake.'

Mrs. Vite she having took her vine,
To Vindsor on they vent to dine:
Ven dinner o'er, Mr. Vite did talk,
'My darling vife, ve'll take a valk:
  The path is vide,
  By vater's side,
So ve vill valk together;
  Vile they gets tea
  For you and me,
Ve vill enjoy the veather.'

Some vanton Eton boys there vere,
Vho mark'd for vaggery this pair:
Mrs. Vite cries out, 'Vat are they ater?'
Ven in they popt Vite in the vater.
    The vicked vits
    Then left the cits,
Ven Vite the vaves sunk under;
    She vept, she squall'd,
    She vail'd, she bawl'd,
'Vill not none help, I vonder.'

Her vimpering vords assistance brought,
Then, vith a boat-hook, Vite they sought;
Ven she, vith expectation big,
Thought Vite was found, but 'twas his vig.
    Vite was not found,
    For he vas drown'd:
To stop her grief each bid her;
    'Ah! no,' she cry'd,
    'I vas a bride,
But now I is a vidor.'

56 *A Maiden There Lived*

A MAIDEN there lived in a large market-town,
Whose skin was much fairer—than any that's brown—
Her eyes were as dark as the coals in the mine,
And when they weren't shut, why they always would
    shine.
  With a black eye, blue eye, blear eye, pig's eye,
    swivel eye, and squinting.

Between her two eyes an excrescence arose,
Which the vulgar call snout, but which I call a nose;
An emblem of sense, it should seem to appear,
For without one we'd look very foolish and queer:
  With your Roman, Grecian, snub-nose, pug-nose,
    snuffling snout, and sneezing.

Good-natured she look'd, that's when out of a frown,
And blush'd like a rose—when the paint was put on;
At church ev'ry morning her prayers she could scan,
And each night sigh and think of—the duty of man,
  With her groaning, moaning, sighing, dying,
    tabernacle—love-feasts.

The follies of youth she had long given o'er,
For the virgin I sing of—was turn'd fifty-four:
Yet suitors she had, who, with words sweet as honey,
Strove hard to possess the bright charms of her money,
  With her household, leasehold, freehold, and her
    copyhold and tenement.

The first who appear'd on this am'rous list,
Was a tailor, who swore by his thimble and twist,
That if his strong passion she e'er should refuse,
He'd depart from the world, shop, cabbage, and goose,
  With his waistcoat, breeches, measures, scissors,
    button-holes, and buckram.

The next was a butcher, of slaughter-ox fame,
A very great boor, and Dick Hog was his name;
He swore she was lamb—but she laugh'd at his pains,
For she hated calf's head—unless served up with brains.
  With his sheep's head, lamb's fry, chitterlins—
    his marrow-bones and cleavers.

After many debates, which occasion'd much strife,
'Mongst love-sick admirers to make her their wife,
To end each dispute came a man out of breath,
Who eloped with the maid, and his name was grim Death.
  With his pick-axe, sexton, coffin, funeral, skeleton,
    and bone-house.

W. M. PRAED

1802–1839

57

Arrivals at a Watering-Place

'I PLAY a spade.—Such strange new faces
  Are flocking in from near and far;
Such frights!—(Miss Dobbs holds all the aces)—
  One can't imagine who they are:
The lodgings at enormous prices,—
  New donkeys, and another fly;
And Madame Bonbon out of ices,
  Although we're scarcely in July:
We're quite as sociable as any,
  But one old horse can scarcely crawl;
And really, where there are so many,
  We can't tell where we ought to call.

'Pray who has seen the odd old fellow
  Who took the Doctor's house last week?—
A pretty chariot,—livery yellow,
  Almost as yellow as his cheek;
A widower, sixty-five, and surly,
  And stiffer than a poplar-tree;
Drinks rum and water, gets up early
  To dip his carcass in the sea;
He's always in a monstrous hurry,
  And always talking of Bengal;
They say his cook makes noble curry;—
  I think, Louisa, we should call.

'And so Miss Jones, the mantua-maker,
  Has let her cottage on the hill!—
The drollest man,—a sugar-baker
  Last year imported from the till;
Prates of his '*orses* and his '*oney*,
  Is quite in love with fields and farms;
A horrid Vandal,—but his money
  Will buy a glorious coat of arms;

Old Clyster makes him take the waters;
  Some say he means to give a ball;
And after all, with thirteen daughters,
  I think, Sir Thomas, you might call.

'That poor young man!—I'm sure and certain
  Despair is making up his shroud;
He walks all night beneath the curtain
  Of the dim sky and murky cloud;
Draws landscapes,—throws such mournful glances;
  Writes verses,—has such splendid eyes;
An ugly name,—but Laura fancies
  He's some great person in disguise!—
And since his dress is all the fashion,
  And since he's very dark and tall,
I think that out of pure compassion,
  I'll get Papa to go and call.

'So Lord St. Ives is occupying
  The whole of Mr. Ford's hotel!
Last Saturday his man was trying
  A little nag I want to sell.
He brought a lady in the carriage;
  Blue eyes,—eighteen, or thereabouts;—
Of course, you know, we *hope* it's marriage,
  But yet the *femme de chambre* doubts.
She looked so pensive when we met her,
  Poor thing!—and such a charming shawl!—
Well! till we understand it better,
  It's quite impossible to call!

'Old Mr. Fund, the London Banker,
  Arrived to-day at Premium Court;
I would not, for the world, cast anchor
  In such a horrid dangerous port;
Such dust and rubbish, lath and plaster,—
  (Contractors play the meanest tricks)—
The roof's as crazy as its master,
  And he was born in fifty-six;
Stairs creaking—cracks in every landing,—
  The colonnade is sure to fall;
We shan't find post or pillar standing,
  Unless we make great haste to call.

'Who was that sweetest of sweet creatures
  Last Sunday in the Rector's seat?
The finest shape,—the loveliest features,—
  I never saw such tiny feet!
My brother,—(this is quite between us)
  Poor Arthur,—'twas a sad affair;
Love at first sight!—she's quite a Venus,
  But then she's poorer far than fair;
And so my father and my mother
  Agreed it would not do at all;
And so,—I'm sorry for my brother!—
  It's settled that we're not to call.

'And there's an author, full of knowledge;
  And there's a captain on half-pay;
And there's a baronet from college,
  Who keeps a boy and rides a bay;
And sweet Sir Marcus from the Shannon,
  Fine specimen of brogue and bone;
And Doctor Calipee, the canon,
  Who weighs, I fancy, twenty stone:
A maiden lady is adorning
  The faded front of Lily Hall:—
Upon my word, the first fine morning,
  We'll make a round, my dear, and call.'

Alas! disturb not, maid and matron,
  The swallow in my humble thatch;
Your son may find a better patron,
  Your niece may meet a richer match:
I can't afford to give a dinner,
  I never was on Almack's list;
And, since I seldom rise a winner,
  I never like to play at whist:
Unknown to me the stocks are falling,
  Unwatched by me the glass may fall:
Let all the world pursue its calling,—
  I'm not at home if people call.

58 *Good Night to the Season*

GOOD night to the Season! 'Tis over!
  Gay dwellings no longer are gay;
The courtier, the gambler, the lover,
  Are scattered like swallows away:
There's nobody left to invite one
  Except my good uncle and spouse;
My mistress is bathing at Brighton,
  My patron is sailing at Cowes:
For want of a better employment,
  Till Ponto and Don can get out,
I'll cultivate rural enjoyment,
  And angle immensely for trout.

Good night to the Season!—the lobbies,
  Their changes, and rumours of change,
Which startled the rustic Sir Bobbies,
  And made all the Bishops look strange;
The breaches, and battles, and blunders,
  Performed by the Commons and Peers;
The Marquis's eloquent blunders,
  The Baronet's eloquent ears;
Denouncings of Papists and treasons,
  Of foreign dominion and oats;
Misrepresentations of reasons,
  And misunderstandings of notes.

Good night to the Season!—the buildings
  Enough to make Inigo sick;
The paintings, and plasterings, and gildings
  Of stucco, and marble, and brick;
The orders deliciously blended,
  From love of effect, into one;
The club-houses only intended,
  The palaces only begun;
The hell, where the fiend in his glory
  Sits staring at putty and stones,
And scrambles from story to story,
  To rattle at midnight his bones.

Good night to the Season!—the dances,
The fillings of hot little rooms,
The glancings of rapturous glances,
The fancyings of fancy costumes;
The pleasures which fashion makes duties,
The praisings of fiddles and flutes,
The luxury of looking at Beauties,
The tedium of talking to mutes;
The female diplomatists, planners
Of matches for Laura and Jane;
The ice of her Ladyship's manners,
The ice of his Lordship's champagne.

Good night to the Season!—the rages
Led off by the chiefs of the throng,
The Lady Matilda's new pages,
The Lady Eliza's new song;
Miss Fennel's macaw, which at Boodle's
Was held to have something to say;
Mrs. Splenetic's musical poodles,
Which bark '*Batti Batti*' all day;
The pony Sir Araby sported,
As hot and as black as a coal,
And the Lion his mother imported,
In bearskins and grease, from the Pole.

Good night to the Season!—the Toso,
So very majestic and tall;
Miss Ayton, whose singing was so-so,
And Pasta, divinest of all;
The labour in vain of the ballet,
So sadly deficient in stars;
The foreigners thronging the Alley,
Exhaling the breath of cigars;
The *loge* where some heiress (how killing!)
Environed with exquisites sits,
The lovely one out of her drilling,
The silly ones out of their wits.

Good night to the Season!—the splendour
  That beamed in the Spanish Bazaar;
Where I purchased—my heart was so tender—
  A card-case, a pasteboard guitar,
A bottle of perfume, a girdle,
  A lithographed Riego, full-grown,
Whom bigotry drew on a hurdle
  That artists might draw him on stone;
A small panorama of Seville,
  A trap for demolishing flies,
A caricature of the Devil,
  And a look from Miss Sheridan's eyes.

Good night to the Season!—the flowers
  Of the grand horticultural fête,
When boudoirs were quitted for bowers,
  And the fashion was—not to be late;
When all who had money and leisure
  Grew rural o'er ices and wines,
All pleasantly toiling for pleasure,
  All hungrily pining for pines,
And making of beautiful speeches,
  And marring of beautiful shows,
And feeding on delicate peaches,
  And treading on delicate toes.

Good night to the Season!—Another
  Will come, with its trifles and toys,
And hurry away, like its brother,
  In sunshine, and odour, and noise.
Will it come with a rose or a briar?
  Will it come with a blessing or curse?
Will its bonnets be lower or higher?
  Will its morals be better or worse?
Will it find me grown thinner or fatter,
  Or fonder of wrong or of right,
Or married—or buried?—no matter:
  Good night to the Season—good night!

59

A Letter of Advice

YOU tell me you're promised a lover,
  My own Araminta, next week;
Why cannot my fancy discover
  The hue of his coat and his cheek?
Alas! if he look like another,
  A vicar, a banker, a beau,
Be deaf to your father and mother,
  My own Araminta, say 'No!'

Miss Lane, at her Temple of Fashion,
  Taught us both how to sing and to speak,
And we loved one another with passion,
  Before we had been there a week:
You gave me a ring for a token;
  I wear it wherever I go;
I gave you a chain,—is it broken?
  My own Araminta, say 'No!'

O think of our favourite cottage,
  And think of our dear Lalla Rookh!
How we shared with the milkmaids their pottage,
  And drank of the stream from the brook;
How fondly our loving lips faltered
  'What further can grandeur bestow?'
My heart is the same;—is yours altered?
  My own Araminta, say 'No!'

Remember the thrilling romances
  We read on the bank in the glen;
Remember the suitors our fancies
  Would picture for both of us then.
They wore the red cross on their shoulder,
  They had vanquished and pardoned their foe—
Sweet friend, are you wiser or colder?
  My own Araminta, say 'No!'

You know, when Lord Rigmarole's carriage
  Drove off with your cousin Justine,
You wept, dearest girl, at the marriage,
  And whispered 'How base she has been!'

You said you were sure it would kill you,
  If ever your husband looked so;
And you will not apostatize,—will you?
  My own Araminta, say 'No!'

When I heard I was going abroad, love,
  I thought I was going to die;
We walked arm in arm to the road, love,
  We looked arm in arm to the sky;
And I said 'When a foreign postilion
  Has hurried me off to the Po,
Forget not Medora Trevilian:
  My own Araminta, say 'No!'

We parted! but sympathy's fetters
  Reach far over valley and hill;
I muse o'er your exquisite letters,
  And feel that your heart is mine still;
And he who would share it with me, love,—
  The richest of treasures below,—
If he's not what Orlando should be, love,
  My own Araminta, say 'No!'

If he wears a top-boot in his wooing,
  If he comes to you riding a cob,
If he talks of his baking or brewing,
  If he puts up his feet on the hob,
If he ever drinks port after dinner,
  If his brow or his breeding is low,
If he calls himself 'Thompson' or 'Skinner,'
  My own Araminta, say 'No!'

If he studies the news in the papers
  While you are preparing the tea,
If he talks of the damps or the vapours
  While moonlight lies soft on the sea,
If he's sleepy while you are capricious,
  If he has not a musical 'Oh!'
If he does not call Werther delicious,—
  My own Araminta, say 'No!'

If he ever sets foot in the City
  Among the stockbrokers and Jews,
If he has not a heart full of pity,
  If he don't stand six feet in his shoes,
If his lips are not redder than roses,
  If his hands are not whiter than snow,
If he has not the model of noses,—
  My own Araminta, say 'No!'

If he speaks of a tax or a duty,
  If he does not look grand on his knees,
If he's blind to a landscape of beauty,
  Hills, valleys, rocks, waters, and trees,
If he dotes not on desolate towers,
  If he likes not to hear the blast blow,
If he knows not the language of flowers,—
  My own Araminta, say 'No!'

He must walk—like a god of old story
  Come down from the home of his rest;
He must smile—like the sun in his glory
  On the buds he loves ever the best;
And oh! from its ivory portal
  Like music his soft speech must flow!—
If he speak, smile, or walk like a mortal,
  My own Araminta, say 'No!'

Don't listen to tales of his bounty,
  Don't hear what they say of his birth,
Don't look at his seat in the county,
  Don't calculate what he is worth;
But give him a theme to write verse on,
  And see if he turns out his toe;
If he's only an excellent person,—
  My own Araminta, say 'No!'

60

Every-Day Characters: Portrait of a Lady

WHAT are you, Lady?—naught is here
  To tell us of your name or story,
To claim the gazer's smile or tear,
  To dub you Whig, or damn you Tory;
It is beyond a poet's skill
  To form the slightest notion, whether
We e'er shall walk through one quadrille,
  Or look upon one moon together.

You're very pretty!—all the world
  Are talking of your bright brow's splendour,
And of your locks, so softly curled,
  And of your hands, so white and slender;
Some think you're blooming in Bengal;
  Some say you're blowing in the city;
Some know you're nobody at all:
  I only feel—you're very pretty.

But bless my heart! it's very wrong;
  You're making all our belles ferocious;
Anne 'never saw a chin so long;'
  And Laura thinks your dress 'atrocious;'
And Lady Jane, who now and then
  Is taken for the village steeple,
Is sure you can't be four feet ten,
  And 'wonders at the taste of people'.

Soon pass the praises of a face;
  Swift fades the very best vermilion;
Fame rides a most prodigious pace;
  Oblivion follows on the pillion;
And all who in these sultry rooms
  Today have stared, and pushed, and fainted,
Will soon forget your pearls and plumes,
  As if they never had been painted.

You'll be forgotten—as old debts
  By persons who are used to borrow;
Forgotten—as the sun that sets,
  When shines a new one on the morrow;
Forgotten—like the luscious peach
  That blessed the schoolboy last September;
Forgotten—like a maiden speech,
  Which all men praise, but none remember.

Yet, ere you sink into the stream
  That whelms alike sage, saint, and martyr,
And soldier's sword, and minstrel's theme,
  And Canning's wit, and Gatton's charter,
Here, of the fortunes of your youth,
  My fancy weaves her dim conjectures,
Which have, perhaps, as much of truth
  As passion's vows, or Cobbett's lectures.

Was't in the north or in the south
  That summer breezes rocked your cradle?
And had you in your baby mouth
  A wooden or a silver ladle?
And was your first unconscious sleep,
  By Brownie banned, or blessed by Fairy?
And did you wake to laugh or weep?
  And were you christened Maud or Mary?

And was your father called 'your grace'?
  And did he bet at Ascot races?
And did he chat of commonplace?
  And did he fill a score of places?
And did your lady-mother's charms
  Consist in picklings, broilings, bastings?
Or did she prate about the arms
  Her brave forefathers wore at Hastings?

Where were you *finished*? tell me where!
  Was it at Chelsea, or at Chiswick?
Had you the ordinary share
  Of books and backboard, harp and physic?
And did they bid you banish pride,
  And mind your Oriental tinting?
And did you learn how Dido died,
  And who found out the art of printing?

And are you fond of lanes and brooks—
  A votary of the sylvan Muses?
Or do you con the little books
  Which Baron Brougham and Vaux diffuses?
Or do you love to knit and sew—
  The fashionable world's Arachne?
Or do you canter down the Row
  Upon a very long-tailed hackney?

And do you love your brother James?
  And do you pet his mares and setters?
And have your friends romantic names?
  And do you write them long long letters?
And are you—since the world began
  All women are—a little spiteful?
And don't you dote on Malibran?
  And don't you think Tom Moore delightful?

I see they've brought you flowers to-day;
  Delicious food for eyes and noses;
But carelessly you turn away
  From all the pinks, and all the roses;
Say, is that fond look sent in search
  Of one whose look as fondly answers?
And is he, fairest, in the Church?
  Or is he—ain't he—in the Lancers?

And is your love a motley page
  Of black and white, half joy, half sorrow?
Are you to wait till you're of age?
  Or are you to be his to-morrow?
Or do they bid you, in their scorn,
  Your pure and sinless flame to smother?
Is he so very meanly born?
  Or are you married to another?

Whate'er you are, at last, adieu!
  I think it is your bounden duty
To let the rhymes I coin for you
  Be prized by all who prize your beauty.
From you I seek nor gold nor fame;
  From you I fear no cruel strictures;
I wish some girls that I could name
  Were half as silent as their pictures!

61

The Talented Man

DEAR Alice! you'll laugh when you know it,—
Last week, at the Duchess's ball,
I danced with the clever new poet,—
You've heard of him,—Tully St. Paul.
Miss Jonquil was perfectly frantic;
I wish you had seen Lady Anne!
It really was very romantic,
He *is* such a talented man!

He came up from Brazenose College,
Just caught, as they call it, this spring;
And his head, love, is stuffed full of knowledge
Of every conceivable thing.
Of science and logic he chatters,
As fine and as fast as he can;
Though I am no judge of such matters,
I'm sure he's a talented man.

His stories and jests are delightful;—
Not stories or jests, dear, for you;
The jests are exceedingly spiteful,
The stories not always *quite* true.
Perhaps to be kind and veracious
May do pretty well at Lausanne;
But it never would answer,—good gracious!
Chez nous—in a talented man.

He sneers,—how my Alice would scold him!—
At the bliss of a sigh or a tear;
He laughed—only think!—when I told him
How we cried o'er Trevelyan last year;
I vow I was quite in a passion;
I broke all the sticks of my fan;
But sentiment's quite out of fashion,
It seems, in a talented man.

Lady Bab, who is terribly moral,
  Has told me that Tully is vain,
And apt—which is silly—to quarrel,
  And fond—which is sad—of champagne.
I listened, and doubted, dear Alice,
  For I saw, when my Lady began,
It was only the Dowager's malice;—
  She *does* hate a talented man!

He's hideous, I own it. But fame, love,
  Is all that these eyes can adore;
He's lame,—but Lord Byron was lame, love,
  And dumpy,—but so is Tom Moore.
Then his voice,—*such* a voice! my sweet creature,
  It's like your Aunt Lucy's toucan:
But oh! what's a tone or a feature,
  When once one's a talented man?

My mother, you know, all the season,
  Has talked of Sir Geoffrey's estate;
And truly, to do the fool reason,
  He *has* been less horrid of late.
But today, when we drive in the carriage,
  I'll tell her to lay down her plan;—
If ever I venture on marriage,
  It must be a talented man!

P.S.—I have found, on reflection,
  One fault in my friend,—*entre nous*;
Without it, he'd just be perfection;—
  Poor fellow, he has not a *sou*!
And so, when he comes in September
  To shoot with my uncle, Sir Dan,
I've promised mamma to remember
  He's *only* a talented man!

LAMAN BLANCHARD

1804–1845

62

Ode to the Human Heart

BLIND Thamyris, and blind Mæonides,
  Pursue the triumph and partake the gale!
Drop tears as fast as the Arabian trees,
  To point a moral or adorn a tale.

Full many a gem of purest ray serene,
  Thoughts that do often lie too deep for tears,
Like angels' visits, few and far between,
  Deck the long vista of departed years.

Man never is, but always to be bless'd;
  The tenth transmitter of a foolish face,
Like Aaron's serpent, swallows up the rest,
  And makes a sunshine in the shady place.

For man the hermit sigh'd, till woman smiled,
  To waft a feather or to drown a fly,
(In wit a man, simplicity a child),
  With silent finger pointing to the sky.

But fools rush in where angels fear to tread,
  Far out amid the melancholy main;
As when a vulture on Imaus bred,
  Dies of a rose in aromatic pain.

Music hath charms to soothe the savage breast,
  Look on her face, and you'll forget them all;
Some mute inglorious Milton here may rest,
  A hero perish, or a sparrow fall.

My way of life is fall'n into the sere;
  I stood in Venice on the Bridge of Sighs,
Like a rich jewel in an Ethiop's ear,
  Who sees through all things with his half-shut eyes.

Oh! for a lodge in some vast wilderness!
  Full many a flower is born to blush unseen,
Fine by degrees and beautifully less,
  And die ere man can say 'Long live the Queen!'

ALFRED, LORD TENNYSON

1809–1892

63

Hendecasyllabics

O YOU chorus of indolent reviewers,
Irresponsible, indolent reviewers,
Look, I come to the test, a tiny poem
All composed in a metre of Catullus,
All in quantity, careful of my motion,
Like the skater on ice that hardly bears him,
Lest I fall unawares before the people,
Waking laughter in indolent reviewers.
Should I flounder awhile without a tumble
Thro' this metrification of Catullus,
They should speak to me not without a welcome,
All that chorus of indolent reviewers.
Hard, hard, hard is it, only not to tumble,
So fantastical is the dainty metre.
Wherefore slight me not wholly, nor believe me
Too presumptuous, indolent reviewers.
O blatant Magazines, regard me rather—
Since I blush to belaud myself a moment—
As some rare little rose, a piece of inmost
Horticultural art, or half coquette-like
Maiden, not to be greeted unbenignly.

64

Somebody

SOMEBODY being a nobody,
Thinking to look like a somebody,
Said that he thought me a nobody:
Good little somebody-nobody,
Had you not known me a somebody,
Would you have called me a nobody?

65

Popular

POPULAR, Popular, Unpopular!
'You're no Poet'—the critics cried!
'Why?' said the Poet. 'You're unpopular!'
Then they cried at the turn of the tide—
'You're no Poet!' 'Why?'—'You're popular!'
Pop-gun, Popular and Unpopular!

OLIVER WENDELL HOLMES

1809–1894

66

Aestivation:
An Unpublished Poem of My Late Latin Tutor

IN candent ire the solar splendour flames;
The foles, languescent, pend from arid rames;
His humid front the cive, anheling, wipes,
And dreams of erring on ventiferous ripes.

How dulce to vive occult to mortal eyes,
Dorm on the herb with none to supervise,
Carp the suave berries from the crescent vine,
And bibe the flow from longicaudate kine!

To me, alas! no verdurous visions come,
Save yon exiguous pool's conferva-scum,—
No concave vast repeats the tender hue
That laves my milk-jug with celestial blue!

Me wretched! Let me curr to quercine shades!
Effund your albid hausts, lactiferous maids!
Oh, might I vole to some umbrageous clump,—
Depart,—be off,—excede,—evade,—erump!

W. M. THACKERAY

1811–1863

67 *Little Billee*

THERE were three sailors in Bristol City,
Who took a boat and went to sea.

But first with beef and captain's biscuit,
And pickled pork they loaded she.

There was guzzling Jack and gorging Jimmy,
And the youngest he was little Bil-*ly*.

Now very soon they were so greedy,
They didn't leave not one split pea.

Says guzzling Jack to gorging Jimmy,
I am confounded hung-*ery*.

Says gorging Jim to guzzling Jacky,
We have no wittles, so we must eat *we*.

Says guzzling Jack to gorging Jimmy,
O gorging Jim, what a fool you be.

There's little Bill as is young and tender,
We're old and tough—so let's eat *he*.

O Bill, we're going to kill and eat you,
So undo the collar of your chemie.

When Bill he heard this information,
He used his pocket-handkerchee.

O let me say my Catechism,
As my poor mammy taught to me.

Make haste, make haste, says guzzling Jacky,
Whilst Jim pulled out his snicker-snee.

So Bill went up the main top-gallant mast,
When down he fell on his bended knee.

He scarce had said his Catechism,
When up he jumps: 'There's land I see!

'There's Jerusalem and Madagascar,
And North and South Ameri-*key*.

'There's the British fleet a-riding at anchor,
With Admiral Napier, K.C.B.'

So when they came to the Admiral's vessel,
He hanged fat Jack, and flogged Jim-*my*.

But as for little Bill, he made him
The Captain of a Seventy-three.

68 *The Sorrows of Werther*

WERTHER had a love for Charlotte
  Such as words could never utter;
Would you know how first he met her?
  She was cutting bread and butter.

Charlotte was a married lady,
  And a moral man was Werther,
And for all the wealth of Indies,
  Would do nothing for to hurt her.

So he sighed and pined and ogled,
  And his passion boiled and bubbled,
Till he blew his silly brains out,
  And no more was by it troubled.

Charlotte, having seen his body
  Borne before her on a shutter,
Like a well-conducted person,
  Went on cutting bread and butter.

EDWARD LEAR

1812–1888

69

How Pleasant to Know . . .

'How pleasant to know Mr. Lear!'
  Who has written such volumes of stuff!
Some think him ill-tempered and queer,
  But a few think him pleasant enough.

His mind is concrete and fastidious,
  His nose is remarkably big;
His visage is more or less hideous,
  His beard it resembles a wig.

He has ears, and two eyes, and ten fingers,
  Leastways if you reckon two thumbs;
Long ago he was one of the singers,
  But now he is one of the dumbs.

He sits in a beautiful parlour,
  With hundreds of books on the wall
He drinks a great deal of Marsala,
  But never gets tipsy at all.

He has many friends, laymen and clerical,
  Old Foss is the name of his cat:
His body is perfectly spherical,
  He weareth a runcible hat.

When he walks in a waterproof white,
  The children run after him so!
Calling out, 'He's come out in his night-
  gown, that crazy old Englishman, oh!'

He weeps by the side of the ocean,
  He weeps on the top of the hill;
He purchases pancakes and lotion,
  And chocolate shrimps from the mill.

He reads but he cannot speak Spanish,
  He cannot abide ginger-beer:
Ere the days of his pilgrimage vanish,
  How pleasant to know Mr. Lear!

70

Limericks

(i)

THERE was an Old Man of Thermopylae,
Who never did anything properly;
But they said, 'If you choose,
To boil eggs in your shoes,
You shall never remain in Thermopylae.'

(ii)

THERE was an Old Man with a beard,
Who said, 'It is just as I feared!—
Two Owls and a Hen,
Four Larks and a Wren,
Have all built their nests in my beard!'

(iii)

THERE was an Old Man who said, 'Hush!
I perceive a young bird in this bush!'
When they said—'Is it small?'
He replied—'Not at all!
It is four times as big as the bush!'

(iv)

THERE was an Old Man of Kamschatka,
Who possessed a remarkably fat cur.
His gait and his waddle,
Were held as a model,
To all the fat dogs in Kamschatka.

ROBERT BROWNING

1812–1889

71

Soliloquy of the Spanish Cloister

GR-R-R—there go, my heart's abhorrence!
  Water your damned flower-pots, do!
If hate killed men, Brother Lawrence,
  God's blood, would not mine kill you!
What? your myrtle-bush wants trimming?
  Oh, that rose has prior claims—
Needs its leaden vase filled brimming?
  Hell dry you up with its flames!

At the meal we sit together:
  Salve tibi! I must hear
Wise talk of the kind of weather,
  Sort of season, time of year:
Not a plenteous cork-crop: scarcely
  Dare we hope oak-galls, I doubt:
What's the Latin name for 'parsley'?
  What's the Greek name for Swine's Snout?

Whew! We'll have our platter burnished,
  Laid with care on our own shelf!
With a fire-new spoon we're furnished,
  And a goblet for ourself,
Rinsed like something sacrificial
  Ere 'tis fit to touch our chaps—
Marked with L. for our initial!
  (He he! There his lily snaps!)

Saint, forsooth! While brown Dolores
  Squats outside the Convent bank
With Sanchicha, telling stories,
  Steeping tresses in the tank,
Blue-black, lustrous, thick like horsehairs,
  —Can't I see his dead eye glow,
Bright as 'twere a Barbary corsair's?
  (That is, if he'd let it show!)

When he finishes refection,
  Knife and fork he never lays
Cross-wise, to my recollection,
  As do I, in Jesu's praise,
I the Trinity illustrate,
  Drinking watered orange-pulp—
In three sips the Arian frustrate;
  While he drains his at one gulp.

Oh, those melons! If he's able
  We're to have a feast; so nice!
One goes to the Abbot's table,
  All of us get each a slice.
How go on your flowers? None double?
  Not one fruit-sort can you spy?
Strange!—And I, too, at such trouble,
  Keep them close-nipped on the sly!

There's a great text in Galatians,
  Once you trip on it, entails
Twenty-nine distinct damnations,
  One sure, if another fails:
If I trip him just a-dying,
  Sure of Heaven as sure can be,
Spin him round and send him flying
  Off to Hell, a Manichee?

Or, my scrofulous French novel
  On grey paper with blunt type!
Simply glance at it, you grovel
  Hand and foot in Belial's gripe:
If I double down its pages
  At the woeful sixteenth print,
When he gathers his greengages,
  Ope a sieve and slip it in't?

Or, there's Satan!—one might venture
  Pledge one's soul to him, yet leave
Such a flaw in the indenture
  As he'd miss till, past retrieve,
Blasted lay that rose-acacia
  We're so proud of! *Hy, Zy, Hine* . . .
'St, there's Vespers! *Plena gratiâ*
  Ave, Virgo! Gr-r-r—you swine!

SHIRLEY BROOKS

1816–1874

72

For A' That

More luck to honest poverty,
  It claims respect, and a' that;
But honest wealth's a better thing,
  We dare be rich for a' that.
    For a' that, and a' that,
      And spooney cant and a' that,
    A man may have a ten-pun note,
      And be a brick for a' that.

What though on soup and fish we dine,
  Wear evening togs and a' that,
A man may like good meat and wine,
  Nor be a knave for a' that.
    For a' that, and a' that,
      Their fustian talk and a' that,
    A gentleman, however clean,
      May have a heart for a' that.

You see yon prater called a Beales,
Who bawls and brays and a' that,
Tho' hundreds cheer his blatant bosh,
He's but a goose for a' that.
For a' that, and a' that,
His Bubblyjocks, and a' that,
A man with twenty grains of sense,
He looks and laughs at a' that.

A prince can make a belted knight,
A marquis, duke, and a' that,
And if the title's earned, all right,
Old England's fond of a' that.
For a' that, and a' that,
Beales' balderdash, and a' that,
A name that tells of service done
Is worth the wear, for a' that.

Then let us pray that come it may
And come it will for a' that,
That common sense may take the place
Of common cant and a' that.
For a' that, and a' that,
Who cackles trash and a' that,
Or be he lord, or be he low,
The man's an ass for a' that.

73 *To Disraeli*

BIG BEN is cracked, we needs must own,
Small Ben is sane, past disputation;
Yet we should like to know whose tone
Is most offensive to the nation.

FREDERICK LOCKER-LAMPSON

1821–1895

74

Our Photograph

SHE played me false, but that's not why
I haven't quite forgiven Di,
Although I've tried:
This curl was hers, so brown, so bright,
She gave it me one blissful night,
And—more beside!

In photo we were grouped together;
She wore the darling hat and feather
That I adore;
In profile by her side I sat
Reading my poetry—but that
She'd heard before.

Why, after all, Di threw me over
I never knew, and can't discover,
Or even guess:
Maybe Smith's lyrics, she decided,
Were sweeter than the sweetest I did—
I acquiesce.

A week before their wedding-day
When Smith was called in haste away
To join the Staff,
Di gave to him, with tearful mien,
Our only photograph. I've seen
That photograph.

I've seen it in Smith's album-book!
Just think! her hat – her tender look,
Are now that brute's!
Before she gave it, off she cut
My body, head and lyrics, but
She was obliged, the little slut,
To leave my boots.

C. G. LELAND

1824–1903

75

Hans Breitmann's Barty

HANS BREITMANN gife a barty;
  Dey had biano-blayin',
I felled in lofe mit a Merican frau,
  Her name vas Madilda Yane.
She hat haar as prown ash a pretzel,
  Her eyes vas himmel-plue,
Und vhen dey looket indo mine,
  Dey shplit mine heart in dwo.

Hans Breitmann gife a barty,
  I vent dere you'll pe pound;
I valtzet mit Madilda Yane,
  Und vent shpinnen' round und round.
De pootiest Fraulein in de house,
  She vayed 'pout dwo hoondred pound,
Und efery dime she gife a shoomp
  She make de vindows sound.

Hans Breitmann gife a barty,
  I dells you it cost him dear;
Dey rolled in more ash sefen kecks
  Of foost-rate lager beer.
Und vhenefer dey knocks de shpicket in
  De Deutschers gifes a cheer;
I dinks dot so vine a barty
  Nefer coom to a het dis year.

Hans Breitmann gife a barty;
  Dere all vas Souse and Brouse,
Vhen de sooper comed in, de gompany
  Did make demselfs to house;
Dey ate das Brot and Gensy broost,
  De Bratwurst and Braten vine,
Und vash der Abendessen down
  Mit four parrels of Neckarwein.

Hans Breitmann gife a barty;
  Ve all cot troonk ash bigs.
I poot mine mout' to a parrel of beer,
  Und emptied it oop mit a schwigs;
Und den I gissed Madilda Yane,
  Und she shlog me on de kop,
Und de gompany vighted mit daple-lecks
  Dill de coonshtable made oos shtop.

Hans Breitmann gife a barty—
  Vhere ish dot barty now?
Vhere ish de lofely golden cloud
  Dot float on de moundain's prow?
Vhere ish de himmelstrahlende stern—
  De shtar of de shpirit's light?
All goned afay mit de lager beer—
  Afay in de ewigkeit!

76 *Ballad by Hans Breitmann*

DER noble Ritter Hugo
  Von Schwillensaufenstein,
Rode out mit shpeer and helmet,
  Und he coom to de panks of de Rhine.

Und oop dere rose a meermaid,
  Vot hadn't got nodings on,
Und she say, 'Oh, Ritter Hugo,
  Where you goes mit yourself alone?'

And he says, 'I rides in de creenwood,
  Mit helmet und mit shpeer,
Till I cooms into ein Gasthaus,
  Und dere I trinks some beer.'

Und den outshpoke de maiden
  Vot hadn't got nodings on:
'I ton't dink mooch of beoplesh
  Dat goes mit demselfs alone.

'You'd petter coom down in de wasser,
 Vhere dere's heaps of dings to see,
Und hafe a shplendid tinner
 Und drafel along mit me.

'Dere you sees de fisch a schwimmin',
 Und you catches dem efery von:'—
So sang dis wasser maiden
 Vot hadn't got nodings on.

'Dere ish drunks all full mit money
 In ships dat vent down of old;
Und you helpsh yourself, by dunder!
 To shimmerin' crowns of gold.

'Shoost look at dese shpoons und vatches!
 Shoost see dese diamant rings!
Coom down and fill your bockets,
 Und I'll giss you like efery dings.

'Vot you vantsh mit your schnapps und lager?
  Coom down into der Rhine!
Der ish pottles der Kaiser Charlemagne
 Vonce filled mit gold-red wine!'

Dat fetched him—he shtood all shpell pound;
 She pooled his coat-tails down,
She drawed him oonder der wasser,
 De maiden mit nodings on.

ANONYMOUS

77

I Saw a Fishpond

I SAW a fishpond all on fire
I saw a house bow to a squire
I saw a parson twelve feet high
I saw a cottage near the sky
I saw a balloon made of lead
I saw a coffin drop down dead
I saw two sparrows run a race
I saw two horses making lace
I saw a girl just like a cat
I saw a kitten wear a hat
I saw a man who saw these too
And said though strange they were all true.

C. S. CALVERLEY

1831–1884

78

Companions,
A Tale of a Grandfather
By the Author of 'Dewy Memories', &c.

I KNOW not of what we ponder'd
  Or made pretty pretence to talk
As, her hand within mine, we wander'd
  Tow'rd the pool by the limetree walk,
While the dew fell in showers from the passion flowers
  And the blush-rose bent on her stalk.

I cannot recall her figure:
  Was it regal as Juno's own?
Or only a trifle bigger
  Than the elves who surround the throne
Of the Faëry Queen, and are seen, I ween,
  By mortals in dreams alone?

What her eyes were like, I know not:
  Perhaps they were blurr'd with tears;
And perhaps in your skies there glow not
  (On the contrary) clearer spheres.
No! as to her eyes I am just as wise
  As you or the cat, my dears.

Her teeth, I presume, were 'pearly':
  But which was she, brunette or blonde?
Her hair, was it quaintly curly,
  Or as straight as a beadle's wand?
That I fail'd to remark;—it was rather dark
  And shadowy round the pond.

Then the hand that reposed so snugly
  In mine—was it plump or spare?
Was the countenance fair or ugly?
  Nay, children, you have me there!
My eyes were p'raps blurr'd; and besides I'd heard
  That it's horribly rude to stare.

And I—was I brusque and surly?
  Or oppressively bland and fond?
Was I partial to rising early?
  Or why did we twain abscond,
All breakfastless too, from the public view
  To prowl by a misty pond?

What pass'd, what was felt or spoken—
  Whether anything pass'd at all—
And whether the heart was broken
  That beat under that shelt'ring shawl—
(If shawl she had on, which I doubt)—has gone,
  Yes, gone from me past recall.

Was I haply the lady's suitor?
  Or her uncle? I can't make out—
Ask your governess, dears, or tutor.
  For myself, I'm in hopeless doubt
As to why we were there, who on earth we were,
  And what this is all about.

79 *'Forever'*

FOREVER: 'tis a single word!
  Our rude forefathers deem'd it two:
Can you imagine so absurd
    A view?

Forever! What abysms of woe
  The word reveals, what frenzy, what
Despair! For ever (printed so)
    Did not.

It looks, ah me! how trite and tame!
  It fails to sadden or appal
Or solace—it is not the same
    At all.

O thou to whom it first occurr'd
  To solder the disjoin'd, and dower
Thy native language with a word
    Of power:

We bless thee! Whether far or near
  Thy dwelling, whether dark or fair
Thy kingly brow, is neither here
    Nor there.

But in men's hearts shall be thy throne,
  While the great pulse of England beats:
Thou coiner of a word unknown
    To Keats!

And nevermore must printer do
  As men did long ago; but run
'For' into 'ever', bidding two
    Be one.

Forever! passion-fraught, it throws
  O'er the dim page a gloom, a glamour:
It's sweet, it's strange; and I suppose
    It's grammar.

Forever! 'Tis a single word!
  And yet our fathers deem'd it two:
Nor am I confident they err'd;
    Are you?

80 *The Schoolmaster Abroad with his Son*

O WHAT harper could worthily harp it,
  Mine Edward! this wide-stretching wold
(Look out *wold*) with its wonderful carpet
  Of emerald, purple, and gold!
Look well at it—also look sharp, it
    Is getting so cold.

The purple is heather (*erica*);
  The yellow, gorse—call'd sometimes 'whin.'
Cruel boys on its prickles might spike a
  Green beetle as if on a pin.
You may roll in it, if you would like a
    Few holes in your skin.

You wouldn't? Then think of how kind you
  Should be to the insects who crave
Your compassion—and then, look behind you
  At yon barley-ears! Don't they look brave
As they undulate (*undulate*, mind you,
    From *unda*, *a wave*).

The noise of those sheep-bells, how faint it
  Sounds here—(on account of our height)!
And this hillock itself—who could paint it,
  With its changes of shadow and light?
Is it not—(never, Eddy, say 'ain't it')—
    A marvellous sight?

Then yon desolate eerie morasses,
  The haunts of the snipe and the hern—
(I shall question the two upper classes
  On *aquatiles*, when we return)—
Why, I see on them absolute masses
    Of *filix* or fern.

How it interests e'en a beginner
(Or *tiro*) like dear little Ned!
Is he listening? As I am a sinner
He's asleep—he is wagging his head.
Wake up! I'll go home to my dinner,
And you to your bed.

The boundless ineffable prairie;
The splendour of mountain and lake
With their hues that seem ever to vary;
The mighty pine-forests which shake
In the wind, and in which the unwary
May tread on a snake;

And this wold with its heathery garment
Are themes undeniably great.
But—although there is not any harm in't—
It's perhaps little good to dilate
On their charms to a dull little varmint
Of seven or eight.

81 *Lines on Hearing the Organ*

GRINDER, who serenely grindest
At my door the Hundredth Psalm,
Till thou ultimately findest
Pence in thy unwashen palm:

Grinder, jocund-hearted Grinder,
Near whom Barbary's nimble son,
Poised with skill upon his hinder
Paws, accepts the proffered bun:

Dearly do I love thy grinding;
Joy to meet thee on thy road
Where thou prowlest through the blinding
Dust with that stupendous load,

'Neath the baleful star of Sirius,
  When the postmen slowlier jog,
And the ox becomes delirious,
  And the muzzle decks the dog.

Tell me by what art thou bindest
  On thy feet those ancient shoon:
Tell me, Grinder, if thou grindest
  Always, always out of tune.

Tell me if, as thou art buckling
  On thy straps with eager claws,
Thou forecastest, inly chuckling,
  All the rage that thou wilt cause.

Tell me if at all thou mindest
  When folks flee, as if on wings,
From thee as at ease thou grindest:
  Tell me fifty thousand things.

Grinder, gentle-hearted Grinder!
  Ruffians who lead evil lives,
Soothed by thy sweet strains, are kinder
  To their bullocks and their wives:

Children, when they see thy supple
  Form approach, are out like shots;
Half-a-bar sets several couple
  Waltzing in convenient spots;

Not with clumsy Jacks or Georges:
  Unprofaned by grasp of man
Maidens speed those simple orgies,
  Betsey Jane with Betsey Ann.

As they love thee in St. Giles's
  Thou art loved in Grosvenor Square:
None of those engaging smiles is
  Unreciprocated there.

Often, ere yet thou hast hammer'd
  Through thy four delicious airs,
Coins are flung thee by enamour'd
  Housemaids upon area stairs:

E'en the ambrosial-whisker'd flunkey
  Eyes thy boots and thine unkempt
Beard and melancholy monkey
  More in pity than contempt.

Far from England, in the sunny
  South, where Anio leaps in foam,
Thou wast rear'd, till lack of money
  Drew thee from thy vineclad home:

And thy mate, the sinewy Jocko,
  From Brazil or Afric came,
Land of simoom and sirocco—
  And he seems extremely tame.

There he quaff'd the undefilèd
  Spring, or hung with apelike glee,
By his teeth or tail or eyelid,
  To the slippery mango-tree:

There he woo'd and won a dusky
  Bride, of instincts like his own;
Talk'd of love till he was husky
  In a tongue to us unknown:

Side by side 'twas theirs to ravage
  The potato ground, or cut
Down the unsuspecting savage
  With the well-aim'd cocoa-nut:—

Till the miscreant Stranger tore him
  Screaming from his blue-faced fair;
And they flung strange raiment o'er him,
  Raiment which he could not bear:

Sever'd from the pure embraces
  Of his children and his spouse,
He must ride fantastic races
  Mounted on reluctant sows:

But the heart of wistful Jocko
  Still was with his ancient flame
In the nutgroves of Morocco;
  Or if not it's all the same.

Grinder, winsome grinsome Grinder!
  They who see thee and whose soul
Melts not at thy charms, are blinder
  Than a trebly-bandaged mole:

They to whom thy curt (yet clever)
  Talk, thy music and thine ape,
Seem not to be joys for ever,
  Are but brutes in human shape.

'Tis not that thy mien is stately,
  'Tis not that thy tones are soft;
'Tis not that I care so greatly
  For the same thing play'd so oft:

But I've heard mankind abuse thee;
  And perhaps it's rather strange,
But I thought that I would choose thee
  For encomium, as a change.

82 *In the Gloaming*

IN the Gloaming to be roaming, where the crested waves are foaming,
  And the shy mermaidens combing locks that ripple to their feet;
When the Gloaming is, I never made the ghost of an endeavour
  To discover—but whatever were the hour, it would be sweet.

'To their feet,' I say, for Leech's sketch indisputably teaches
  That the mermaids of our beaches do not end in ugly tails,
Nor have homes among the corals; but are shod with neat balmorals,
  An arrangement no one quarrels with, as many might with scales.

Sweet to roam beneath a shady cliff, of course with some young lady,
Lalage, Neæra, Haidee, or Elaine, or Mary Ann:
Love, you dear delusive dream, you! Very sweet your victims deem you,
When, heard only by the seamew, they talk all the stuff one can.

Sweet to haste, a licensed lover, to Miss Pinkerton the glover,
Having managed to discover what is dear Neæra's 'size':
P'raps to touch that wrist so slender, as your tiny gift you tender,
And to read you're no offender, in those laughing hazel eyes.

Then to hear her call you 'Harry,' when she makes you fetch and carry—
O young men about to marry, what a blessed thing it is!
To be photograph'd—together—cased in pretty Russian leather—
Hear her gravely doubting whether they have spoilt your honest phiz!

Then to bring your plighted fair one first a ring—a rich and rare one—
Next a bracelet, if she'll wear one, and a heap of things beside;
And serenely bending o'er her, to inquire if it would bore her
To say when her own adorer may aspire to call her bride!

Then, the days of courtship over, with your WIFE to start for Dover
Or Dieppe—and live in clover evermore, whate'er befalls:
For I've read in many a novel that, unless they've souls that grovel,
Folks *prefer* in fact a hovel to your dreary marble halls:

To sit, happy married lovers; Phillis trifling with a plover's
Egg, while Corydon uncovers with a grace the Sally Lunn,
Or dissects the lucky pheasant—that, I think, were passing pleasant;
As I sit alone at present, dreaming darkly of a Dun.

83 FROM *Dover to Munich*

FAREWELL, farewell! Before our prow
Leaps in white foam the noisy channel;
A tourist's cap is on my brow,
My legs are cased in tourist's flannel:

Around me gasp the invalids—
  The quantity to-night is fearful—
I take a brace or so of weeds,
  And feel (as yet) extremely cheerful.

The night wears on:—my thirst I quench
  With one imperial pint of porter;
Then drop upon a casual bench—
  (The bench is short, but I am shorter)—

Place 'neath my head the *havre-sac*
  Which I have stowed my little all in,
And sleep, though moist about the back,
  Serenely in an old tarpaulin.

Bed at Ostend at 5 A.M.
  Breakfast at 6, and train 6·30,
Tickets to Königswinter (mem.
  The seats unutterably dirty).

And onward thro' those dreary flats
  We move, and scanty space to sit on,
Flanked by stout girls with steeple hats,
  And waists that paralyse a Briton;—

By many a tidy little town,
  Where tidy little Fraus sit knitting;
(The men's pursuits are, lying down,
  Smoking perennial pipes, and spitting;)

And doze, and execrate the heat,
  And wonder how far off Cologne is,
And if we shall get aught to eat,
Till we get there, save raw polonies:

Until at last the 'gray old pile'
  Is seen, is past, and three hours later
We're ordering steaks, and talking vile
  Mock-German to an Austrian waiter.

Königswinter, hateful Königswinter!
  Burying-place of all I loved so well!
Never did the most extensive printer
  Print a tale so dark as thou could'st tell!

In the sapphire West the eve yet lingered,
  Bathed in kindly light those hill-tops cold;
Fringed each cloud, and, stooping rosy-fingered,
  Changed Rhine's waters into molten gold:—

While still nearer did his light waves splinter
  Into silvery shafts the streaming light;
And I said I loved thee, Königswinter,
  For the glory that was thine that night.

And we gazed, till slowly disappearing,
  Like a day-dream, passed the pageant by,
And I saw but those lone hills, uprearing
  Dull dark shapes against a hueless sky.

Then I turned, and on those bright hopes pondered
  Whereof yon gay fancies were the type;
And my hand mechanically wandered
  Towards my left-hand pocket for a pipe.

Ah! why starts each eyeball from its socket,
  As, in Hamlet, start the guilty Queen's?
There, deep-hid in its accustomed pocket,
  Lay my sole pipe, smashed to smithereens!

On, on the vessel steals;
Round go the paddle-wheels,
And now the tourist feels
    As he should;
For king-like rolls the Rhine,
And the scenery's divine,
And the victuals and the wine
    Rather good.

From every crag we pass'll
Rise up some hoar old castle;
The hanging fir-groves tassel
    Every slope;
And the vine her lithe arm stretches
Over peasants singing catches—
And you'll make no end of sketches,
    I should hope.

We've a nun here (called Therèse),
Two couriers out of place,
One Yankee with a face
    Like a ferret's:
And three youths in scarlet caps
Drinking chocolate and schnapps—
A diet which perhaps
    Has its merits.

And day again declines:
In shadow sleep the vines,
And the last ray thro' the pines
    Feebly glows,
Then sinks behind yon ridge;
And the usual evening midge
Is settling on the bridge
    Of my nose.

And keen's the air and cold,
And the sheep are in the fold,
And Night walks sable-stoled
    Thro' the trees;
And on the silent river
The floating starbeams quiver;—
And now, the saints deliver
    Us from fleas.

LEWIS CARROLL

1832–1898

84 *How Doth the Little Crocodile*

How doth the little crocodile
  Improve his shining tail,
And pour the waters of the Nile
  On every golden scale!

How cheerfully he seems to grin
  How neatly spreads his claws,
And welcomes little fishes in,
  With gently smiling jaws!

85 *'You are old, Father William'*

'You are old, Father William,' the young man said,
  'And your hair has become very white;
And yet you incessantly stand on your head—
  Do you think, at your age, it is right?'

'In my youth,' Father William replied to his son,
  'I feared it might injure the brain;
But, now that I'm perfectly sure I have none,
  Why, I do it again and again.'

'You are old,' said the youth, 'as I mentioned before,
  And have grown most uncommonly fat;
Yet you turned a back-somersault in at the door—
  Pray, what is the reason of that?'

'In my youth,' said the sage, as he shook his grey locks,
  'I kept all my limbs very supple
By the use of this ointment—one shilling the box—
  Allow me to sell you a couple?'

'You are old,' said the youth, 'and your jaws are too weak
  For anything tougher than suet;
Yet you finished the goose, with the bones and the beak—
  Pray, how did you manage to do it?'

'In my youth,' said his father, 'I took to the law,
  And argued each case with my wife;
And the muscular strength, which it gave to my jaw,
  Has lasted the rest of my life.'

'You are old,' said the youth, 'one would hardly suppose
  That your eye was as steady as ever;
Yet you balance an eel on the end of your nose—
  What made you so awfully clever?'

'I have answered three questions, and that is enough,'
  Said his father. 'Don't give yourself airs!
Do you think I can listen all day to such stuff?
  Be off, or I'll kick you down-stairs!'

86 *'Twinkle, twinkle, little bat!'*

TWINKLE, twinkle, little bat!
How I wonder what you're at!
Up above the world you fly,
Like a tea-tray in the sky.

87 *'Tis the Voice of the Lobster*

(i)

' 'TIS the voice of the Lobster: I heard him declare
"You have baked me too brown, I must sugar my hair."
As a duck with its eyelids, so he with his nose
Trims his belt and his buttons, and turns out his toes.
When the sands are all dry, he is gay as a lark,
And will talk in contemptuous tones of the Shark:
But, when the tide rises and sharks are around,
His voice has a timid and tremulous sound.'

(ii)

I passed by his garden, and marked, with one eye,
How the Owl and the Panther were sharing a pie:
The Panther took pie-crust, and gravy, and meat,
While the Owl had the dish as its share of the treat.
When the pie was all finished, the Owl, as a boon,
Was kindly permitted to pocket the spoon:
While the Panther received knife and fork with a growl.
And concluded the banquet by——

88

Jabberwocky

'TWAS brillig, and the slithy toves
  Did gyre and gimble in the wabe;
All mimsy were the borogoves,
  And the mome raths outgrabe.

'Beware the Jabberwock, my son!
  The jaws that bite, the claws that catch!
Beware the Jubjub bird, and shun
  The frumious Bandersnatch!'

He took his vorpal sword in hand:
  Long time the manxome foe he sought—
So rested he by the Tumtum tree,
  And stood awhile in thought.

And as in uffish thought he stood,
  The Jabberwock, with eyes of flame,
Came whiffling through the tulgey wood,
  And burbled as it came!

One, two! One, two! And through and through
  The vorpal blade went snicker-snack!
He left it dead, and with its head
  He went galumphing back.

'And hast thou slain the Jabberwock!
  Come to my arms, my beamish boy!
O frabjous day! Callooh! Callay!'
  He chortled in his joy.

'Twas brillig, and the slithy toves
  Did gyre and gimble in the wabe;
All mimsy were the borogoves,
  And the mome raths outgrabe.

89 *The Walrus and the Carpenter*

THE sun was shining on the sea,
  Shining with all his might:
He did his very best to make
  The billows smooth and bright—
And this was odd, because it was
  The middle of the night.

The moon was shining sulkily,
  Because she thought the sun
Had got no business to be there
  After the day was done—
'It's very rude of him,' she said,
  'To come and spoil the fun!'

The sea was wet as wet could be,
  The sands were dry as dry.
You could not see a cloud, because
  No cloud was in the sky:
No birds were flying overhead—
  There were no birds to fly.

The Walrus and the Carpenter
  Were walking close at hand:
They wept like anything to see
  Such quantities of sand:
'If this were only cleared away,'
  They said, 'it *would* be grand!'

'If seven maids with seven mops
  Swept it for half a year,
Do you suppose,' the Walrus said,
  'That they could get it clear?'
'I doubt it,' said the Carpenter,
  And shed a bitter tear.

'O Oysters, come and walk with us!'
  The Walrus did beseech.
'A pleasant walk, a pleasant talk,
  Along the briny beach:
We cannot do with more than four,
  To give a hand to each.'

The eldest Oyster looked at him,
  But not a word he said:
The eldest Oyster winked his eye,
  And shook his heavy head—
Meaning to say he did not choose
  To leave the oyster-bed.

But four young Oysters hurried up,
  All eager for the treat:
Their coats were brushed, their faces washed
  Their shoes were clean and neat—
And this was odd, because, you know,
  They hadn't any feet.

Four other Oysters followed them,
  And yet another four;
And thick and fast they came at last,
  And more, and more, and more—
All hopping through the frothy waves,
  And scrambling to the shore.

The Walrus and the Carpenter
  Walked on a mile or so,
And then they rested on a rock
  Conveniently low:
And all the little Oysters stood
  And waited in a row.

'The time has come,' the Walrus said,
  'To talk of many things:
Of shoes—and ships—and sealing wax—
  Of cabbages—and kings—
And why the sea is boiling hot—
  And whether pigs have wings.'

'But wait a bit,' the Oysters cried,
  'Before we have our chat;
For some of us are out of breath,
  And all of us are fat!'
'No hurry!' said the Carpenter.
  They thanked him much for that.

'A loaf of bread,' the Walrus said,
  'Is what we chiefly need:
Pepper and vinegar besides
  Are very good indeed—
Now, if you're ready, Oysters dear,
  We can begin to feed.'

'But not on us!' the Oysters cried,
  Turning a little blue.
'After such kindness that would be
  A dismal thing to do!'
'The night is fine,' the Walrus said,
  'Do you admire the view?

'It was so kind of you to come,
  And you are very nice!'
The Carpenter said nothing but
  'Cut us another slice.
I wish you were not quite so deaf—
  I've had to ask you twice!'

'It seems a shame,' the Walrus said,
  'To play them such a trick.
After we've brought them out so far,
  And made them trot so quick!'
The Carpenter said nothing but
  'The butter's spread too thick!'

'I weep for you,' the Walrus said:
  'I deeply sympathize.'
With sobs and tears he sorted out
  Those of the largest size,
Holding his pocket-handkerchief
  Before his streaming eyes.

'O Oysters,' said the Carpenter,
  'You've had a pleasant run!
Shall we be trotting home again?'
  But answer came there none—
And this was scarcely odd, because
  They'd eaten every one.

90 *The White Knight's Song*

I'LL tell thee everything I can;
  There's little to relate.
I saw an aged aged man,
  A-sitting on a gate.
'Who are you, aged man?' I said.
  'And how is it you live?'
And his answer trickled through my head
  Like water through a sieve.

He said 'I look for butterflies
  That sleep among the wheat:
I make them into mutton-pies,
  And sell them in the street.
I sell them unto men,' he said,
  'Who sail on stormy seas;
And that's the way I get my bread—
  A trifle, if you please.'

But I was thinking of a plan
  To dye one's whiskers green,
And always use so large a fan
  That they could not be seen.
So, having no reply to give
  To what the old man said,
I cried 'Come, tell me how you live!'
  And thumped him on the head.

His accents mild took up the tale:
  He said 'I go my ways,
And when I find a mountain-rill,
  I set it in a blaze;
And thence they make a stuff they call
  Rowland's Macassar-Oil—
Yet twopence-halfpenny is all
  They give me for my toil.'

But I was thinking of a way
  To feed oneself on batter,
And so go on from day to day
  Getting a little fatter.
I shook him well from side to side,
  Until his face was blue:
'Come, tell me how you live,' I cried,
  'And what it is you do!'

He said 'I hunt for haddocks' eyes
  Among the heather bright,
And work them into waistcoat-buttons
  In the silent night.
And these I do not sell for gold
  Or coin of silvery shine,
But for a copper halfpenny,
  And that will purchase nine.

'I sometimes dig for buttered rolls,
 Or set limed twigs for crabs;
I sometimes search the grassy knolls
 For wheels of Hansom-cabs.
And that's the way' (he gave a wink)
 'By which I get my wealth—
And very gladly will I drink
 Your Honour's noble health.'

I heard him then, for I had just
 Completed my design
To keep the Menai bridge from rust
 By boiling it in wine.
I thanked him much for telling me
 The way he got his wealth,
But chiefly for his wish that he
 Might drink my noble health.

And now, if e'er by chance I put
 My fingers into glue,
Or madly squeeze a right-hand foot
 Into a left-hand shoe,
Or if I drop upon my toe
 A very heavy weight,
I weep, for it reminds me so
Of that old man I used to know—
Whose look was mild, whose speech was slow,
Whose hair was whiter than the snow,
Whose face was very like a crow,
With eyes, like cinders, all aglow,
Who seemed distracted with his woe,
Who rocked his body to and fro,
And muttered mumblingly and low,
As if his mouth were full of dough,
Who snorted like a buffalo—
That summer evening long ago
 A-sitting on a gate.

91 *Hiawatha's Photographing*

FROM his shoulder Hiawatha
Took the camera of rosewood—
Made of sliding, folding rosewood—
Neatly put it all together.
In its case it lay compactly,
Folded into nearly nothing;
But he opened out its hinges,
Pushed and pulled the joints and hinges
Till it looked all squares and oblongs,
Like a complicated figure
In the second book of Euclid.
This he perched upon a tripod,
And the family, in order
Sat before him for their pictures—
Mystic, awful, was the process.
First, a piece of glass he coated
With collodion, and plunged it
In a bath of lunar caustic
Carefully dissolved in water—
There he left it certain minutes.
Secondly, my Hiawatha
Made with cunning hand a mixture
Of the acid pyrro-gallic,
And of glacial-acetic,
And of alcohol and water—
This developed all the picture.
Finally, he fixed each picture
With a saturate solution
Which was made of hyposulphite
Which, again, was made of soda.
(Very difficult the name is
For a metre like the present
But periphrasis has done it.)
All the family, in order,
Sat before him for their pictures;
Each in turn, as he was taken,
Volunteered his own suggestions—
His invaluable suggestions.
First, the governor—the father—
He suggested velvet curtains

Looped about a massy pillar,
And the corner of a table—
Of a rosewood dining table.
He would hold a scroll of something—
Hold it firmly in his left hand;
He would keep his right hand buried
(Like Napoleon) in his waistcoat;
He would gaze upon the distance—
(Like a poet seeing visions,
Like a man that plots a poem,
In a dressing gown of damask,
At 12.30 in the morning,
Ere the servants bring in luncheon)—
With a look of pensive meaning,
As of ducks that die in tempests.
    Grand, heroic was the notion:
Yet the picture failed entirely,
Failed because he moved a little—
Moved because he couldn't help it.
    Next his better half took courage—
She would have her picture taken:
She came dressed beyond description,
Dressed in jewels and in satin,
Far too gorgeous for an empress.
Gracefully she sat down sideways,
With a simper scarcely human,
Holding in her hand a nosegay
Rather larger than a cabbage.
    All the while that she was taking,
Still the lady chattered, chattered,
Like a monkey in the forest.
'Am I sitting still?' she asked him;
'Is my face enough in profile?
Shall I hold the nosegay higher?
Will it come into the picture?'
And the picture failed completely.
    Next the son, the stunning Cantab——
He suggested curves of beauty,
Curves pervading all his figure,
Which the eye might follow onward
Till it centred in the breast-pin—
Centred in the golden breast-pin.
He had learnt it all from Ruskin,

(Author of *The Stones of Venice*,
Seven Lamps of Architecture,
Modern Painters, and some others)—
And perhaps he had not fully
Understood the author's meaning;
But, whatever was the reason,
All was fruitless, as the picture
Ended in a total failure.
    After him the eldest daughter:
She suggested very little,
Only begged she might be taken
With her look of 'passive beauty'.
Her idea of passive beauty
Was a squinting of the left eye,
Was a drooping of the right eye,
Was a smile that went up sideways
To the corner of the nostrils.
    Hiawatha, when she asked him,
Took no notice of the question,
Looked as if he hadn't heard it;
But, when pointedly appealed to,
Smiled in a peculiar manner,
Coughed, and said it 'didn't matter',
Bit his lips, and changed the subject.
    Nor in this was he mistaken,
As the picture failed completely.
    So, in turn, the other daughters:
All of them agreed on one thing,
That their pictures came to nothing,
Though they differed in their causes,
From the eldest, Grinny-haha,
Who, throughout her time of taking,
Shook with sudden, ceaseless laughter,
With a silent fit of laughter,
To the youngest, Dinny-wawa,
Shook with sudden, causeless weeping—
Anything but silent weeping:
And their pictures failed completely.
Last, the youngest son was taken:
'John' his Christian name had once been;
But his overbearing sisters
Called him names he disapproved of—
Called him Johnny, 'Daddy's Darling'—

Called him Jacky, 'Scrubby Schoolboy'.
Very rough and thick his hair was,
Very dusty was his jacket,
Very fidgetty his manner,
And, so fearful was the picture,
In comparison the others
Might be thought to have succeeded—
To have partially succeeded.
    Finally, my Hiawatha
Tumbled all the tribe together
('Grouped' is not the right expression),
And, as happy chance would have it,
Did at last obtain a picture
Where the faces all succeeded:
Each came out a perfect likeness.
    Then they joined and all abused it—
Unrestrainedly abused it—
As 'the worst and ugliest picture
That could possibly be taken
Giving one such strange expressions!
Sulkiness, conceit, and meanness!
Really any one would take us
(Any one who did not know us)
For the most unpleasant people!'
(Hiawatha seemed to think so—
Seemed to think it not unlikely).
All together rang their voices—
Angry, hard, discordant voices—
As of dogs that howl in concert,
As of cats that wail in chorus.
    But my Hiawatha's patience,
His politeness and his patience,
Unaccountably had vanished,
And he left that happy party.
Neither did he leave them slowly,
With the calm deliberation,
The intense deliberation
Of a photographic artist:
But he left them in a hurry,
Left them in a mighty hurry,
Stating that he would not stand it,
Stating in emphatic language
What he'd be before he'd stand it.

Hurriedly he packed his boxes:
Hurriedly the porter trundled
On a barrow all his boxes:
Hurriedly he took his ticket:
Hurriedly the train received him:
Thus departed Hiawatha.

H. J. BYRON

1834–1884

92

Rural Simplicity

SAID I, Oh, give me simplicity,
  Purity, truthful rurality:
Wearied alas, I'm of this city,
  With its deceit and venality.
Give me a home amidst villagers,
  Noted for honest veracity,
Far from this town full of pillagers,
  Reckless in daring audacity.

Long 'twas before the discovery
  Came of the spot I was wishing for:
There in my cottage ('The Dovery')
  Trout though diurnally fishing for,
Fish never ventured a bite at all,
  Though the inhabitants said they would,
Never appeared to my sight at all;
  'P'rhaps if I threw in some bread they would.'

Roger, a humble parishioner,
  Noted for cringing humility,
For the groom's place a petitioner,
  Robbed me with wondrous ability:
Jane, thought a girl of sobriety,
  Praised by our excellent Vicar, and
Timothy, noted for piety,
  Took my loose cash and my liquor, and

Though I induced was to pardon her,
  Soon I regretted my lenity,
For she eloped with the gardener;
  This quite upset my serenity.
Still I had faith in simplicity;
  Giles, too, the pride of the peasantry,
Till I found out that a visit he
  Paid to the eggs in the pheasantry.

Then I'd a Bailiff, a denizen,
  Native of where I had settled at:
Mutton *I* ate, *he* ate venison,
  Which I must own I was nettled at.
I kept a small modest vinery,
  He, in a manner Quixotical,
Went in for peaches and pinery,
  Plants too most rare and exotical.

Famed were his wines and his cookery,
  Famed too his wide hospitality;
My house was rather a rookery
  Since I'd gone in for 'Rurality.'
Though for the fields I'd selected a
  Party of pipers and taborers
Picturesque, still I'd expected a
  Lot who were decentish labourers.

'Sing?' yes, they sung, but their ditties were
  Not to be lauded extensively:
Songs I have heard howled in cities were
  Seldom couched *quite* so offensively.
'Dance?' yes, they danced, but proclivities
  Which have been censured quite recently,
During the gloaming's festivities,
    Peasants developed indecently.

Shocked at this base immorality,
  Sickened with Chloes and Phyllises,
Welcome's the modest formality
  At a big ball,—say at 'Willis's.'
Tytyrus? Corydon?—Bosh!
  Pretty in print, but reality
Proves that the *print* doesn't wash,
  Being inferior quality.
Give me this city of crime
  Seething with vile illegality,
Rather than passing my time
  Searching for 'Simple Rurality.'

GODFREY TURNER

fl. 1878

93 *Tattle*

A SCANDAL or two
I shall mention to you;
And first, here is one
That's as sure as a gun,
For the person who told me was told it
By some one who knew
That the story was true,
For he heard it one day,
In a casual way,
From a man who's a brother
Of some one or other
Who holds an appointment
In Holloway's ointment
And pill branch at Bow;
But, besides that, I *know*.
Well, the Duchess of Dash,
Being troubled for cash,
Pawned an emerald cluster
And large knuckle-duster
Of brilliants and rubies
At Spoonbill and Booby's;
And now they allege
That the forfeited pledge

Is nothing but paste
Set with very good taste,
Which they found was the fact
When the cluster was cracked
By a fall from the 'kerridge'
Of Mrs. Sam Gerridge,
To whom S. and B. had just sold it.
So what do you think of your Duchess of Dash?
With my raggery waggery, alumny calumny, slippery sloppery slash!

You've heard what they say
Of your friend, little K.?
Well, there's a fine mess!
And it can't be much less
Than a thousand he's dropped
In the bank that has stopped;
At a time, I'll be bound,
When he wants every pound;
And I'd not be surprised,
As at present advised,
If we chanced to hear tell
That he'd bolted this minute,
And Johnson as well,
For I know he was in it!
And then there's young Q.,
He'll begin to look blue
When B. and C. find
That there's something behind,
As they're sure before long to discover.
Of course you're aware
Of the Whiffer affair?
I'm an old friend of Whiffer,
And that's why we differ,
And when he was single
I told him my mind
About Emily Bingle,
But he was so blind
That he never could see
She was sweet upon *me*.
Oh! The woman's awake;
Don't make any mistake!
She's a nice-looking girl,
And the niece of an Earl,

But she hadn't a rap
When she married this chap,
And now he begins
To cry 'Needles and pins!'
At least I suppose,
Although nobody knows,
That he must have found out
What his wife is about,
And a certain Lord F. is her lover.
So don't you consider old Whiffer was rash?
With my staggery swaggery, blackery quackery, blundery Grundy-ry hash!

That's rather a queer
Sort of story I hear
Set about to explain
Why Sir John goes to Spain,
And his family too,
With, of course, little Loo.
Yes, it possibly may
Have been just as you say;
But it does appear strange
They should suddenly change
The tour they had planned,
For I can't understand
Why it suits them to go
Where they're sure not to see
A soul whom they know.
Well, it's nothing to me!
But I hear some odd things
That are told of the Byngs;
And one of 'em is
That they're too fond of 'fiz,'
To say nothing of brandy
Or anything handy.
Well, now, don't you think
When a girl takes to drink,
It's all up with her quite
As regards what is right?
There is only one answer I *can* see.
But, talking of that,
Come, I'll bet you a hat

You don't know the reason
Why all through last season
The Byngs gave no ball,
Nor the least thing at all
In the house-warming way.
Oh! I know what you'll say;
They had lost a relation;
But that's mere evasion.
The true cause I know,
But you won't let it go
Any further, I'm sure,
For it's well-known that you're
Not the fellow to chatter
About such a matter;
And so, it's just this,
There was something amiss
That had reached Sir John's ears
To awaken his fears
About Miss Louisa
  And John Thomas Coombe.
(I need not say he's a
  Mere sort of head-groom.)
And John had declared
That if ever she dared
Show herself at a 'hop,'
He'd be there to say 'stop,'
And John would have done it, I fancy.
——Bah! Tattlers, beware of the cudgel or lash,
With your blundery Grundy-ry, blackery quackery, staggery swaggery, slippery sloppery, alumny calumny, raggery waggery, uttery guttery trash!

GODFREY TURNER

94 *The Journal of Society*

OH, have you seen the *Tattlesnake* and have you read what's in it,
About—dear me! Tut! What's his name? I'll tell you in a minute.
Thwaites? No, not Thwaites. McCorquodale? No. Gillingwater?
Coutts?
The man who wears a high-crowned hat and patent-leather boots.
In philanthropic circles he is known as 'Ready Bob';
I think he lives out Kilburn way and rides a chestnut cob.
He gets it in the *Tattlesnake!* I wonder how he feels
When all his friends their shoulders shrug and turn upon their heels.
We thought him such a thoroughbred! And now it's all found out.
They'd hardly dare to write like that, if there were any doubt.
Besides, 'twixt you and me, you know, there *was* a something queer
About his early history, I think I used to hear.
You don't remember? Well, I do; although I could not say
Precisely what the story was, at such a distant day.
These things will always hang about a fellow's after-life.
Oh, now I know! Yes. Was there not some talk about his wife?
Or else it was his sister, or his mother, or his aunt:
Though if I'm asked to give the facts, I frankly say I can't.
But get the *Tattlesnake*, my boy; you'll find it worth the money,
Unless, that is, you're one of those who don't think scandal funny.
That class of writing lacks, I own, the literary zeal
To add a charm to Addison, or polish put on Steele.
Between the *Tattlesnake* and *them* it's not a case of choosing;
But personality, if dull, is in its way amusing;
Although the way's not that of wit, nor graciousness, nor grammar.
You would not have a rapier-point upon a blacksmith's hammer!
But what of that? On certain ears wit blunted tells the best,
And satire glads the public heart when as a libel dressed.
Of course I don't defend the thing: it's bad in many ways;
But is not this the 'golden' age? And I suppose it pays.
So get the *Tattlesnake*.—It's dead? Bless me! You don't mean that?
Well, after all, I'm not surprised. I thought it rather flat.
By chance I saw it once or twice, and then I found it slow;
That shameful article I read about a month ago.
But still some dirt will always stick; and, long as he may live,
That that same dirt had not been flung a thousand pounds he'd give;—
When all his old acquaintance cut and every urchin hoots
The man who wears the high-crowned hat and patent-leather boots.

W. S. GILBERT

1836–1911

95 *The Darned Mounseer*

I SHIPPED, d'ye see, in a Revenue sloop,
And, off Cape Finisteere,
A merchantman we see,
A Frenchman, going free,
So we made for the bold Mounseer,
D'ye see?
We made for the bold Mounseer!
But she proved to be a Frigate—and she up with her ports,
And fires with a thirty-two!
It come uncommon near,
But we answered with a cheer,
Which paralysed the Parley-voo,
D'ye see?
Which paralysed the Parley-voo!

Then our Captain he up and he says, says he,
'That chap we need not fear,—
We can take her, if we like,
She is sartin for to strike,
For she's only a darned Mounseer,
D'ye see?
She's only a darned Mounseer!
But to fight a French fal-lal it's like hittin' of a gal—
It's a lubberly thing for to do;
For we, with all our faults,
Why, we're sturdy British salts,
While she's but a Parley-voo,
D'ye see?
A miserable Parley-voo!'

So we up with our helm, and we scuds before the breeze,
As we gives a compassionating cheer;
Froggee answers with a shout
As he sees us go about,
Which was grateful of the poor Mounseer,
D'ye see?
Which was grateful of the poor Mounseer!
And I'll wager in their joy they kissed each other's cheek
(Which is what them furriners do),
And they blessed their lucky stars
We were hardy British tars
Who had pity on a poor Parley-voo,
D'ye see?
Who had pity on a poor Parley-voo!

96 *The Englishman*

He is an Englishman!
For he himself has said it,
And it's greatly to his credit,
That he is an Englishman!
For he might have been a Roosian,
A French, or Turk, or Proosian,
Or perhaps Itali-an!
But in spite of all temptations,
To belong to other nations,
He remains an Englishman!
Hurrah!
For the true-born Englishman!

W. S. GILBERT

97 *The Modern Major-General*

I AM the very pattern of a modern Major-Gineral,
I've information vegetable, animal, and mineral;
I know the kings of England, and I quote the fights historical,
From Marathon to Waterloo, in order categorical;
I'm very well acquainted, too, with matters mathematical,
I understand equations, both the simple and quadratical;
About binomial theorem I'm teeming with a lot o' news,
With interesting facts about the square of the hypotenuse.
I'm very good at integral and differential calculus,
I know the scientific names of beings animalculous.
In short, in matters vegetable, animal, and mineral,
I am the very model of a modern Major-Gineral.

I know our mythic history—KING ARTHUR'S and SIR CARADOC'S,
I answer hard acrostics, I've a pretty taste for paradox;
I quote in elegiacs all the crimes of HELIOGABALUS,
In conics I can floor peculiarities parabolous.
I tell undoubted RAPHAELS from GERARD DOWS and ZOFFANIES,
I know the croaking chorus from the 'Frogs' of ARISTOPHANES;
Then I can hum a fugue, of which I've heard the music's din afore,
And whistle all the airs from that confounded nonsense 'Pinafore.'
Then I can write a washing-bill in Babylonic cuneiform,
And tell you every detail of CARACTACUS'S uniform.
In short, in matters vegetable, animal, and mineral,
I am the very model of a modern Major-Gineral.

In fact, when I know what is meant by 'mamelon' and 'ravelin,'
When I can tell at sight a Chassepôt rifle from a javelin,
When such affairs as *sorties* and surprises I'm more wary at,
And when I know precisely what is meant by Commissariat,
When I have learnt what progress has been made in modern gunnery,
When I know more of tactics than a novice in a nunnery,
In short, when I've a smattering of elementary strategy,
You'll say a better Major-Giner*al* has never *sat* a gee—
For my military knowledge, though I'm plucky and adventury,
Has only been brought down to the beginning of the century.
But still in learning vegetable, animal, and mineral,
I am the very model of a modern Major-Gineral!

98

The Policeman's Lot

WHEN a felon's not engaged in his employment
  Or maturing his felonious little plans,
His capacity for innocent enjoyment
  Is just as great as any honest man's.
Our feelings we with difficulty smother
  When constabulary duty's to be done:
Ah, take one consideration with another,
  A policeman's lot is not a happy one!

When the enterprising burglar isn't burgling,
  When the cut-throat isn't occupied in crime,
He loves to hear the little brook a-gurgling,
  And listen to the merry village chime.
When the coster's finished jumping on his mother,
  He loves to lie a-basking in the sun:
Ah, take one consideration with another,
  The policeman's lot is not a happy one!

99

The Yarn of the 'Nancy Bell'

'TWAS on the shores that round our coast
  From Deal to Ramsgate span,
That I found alone on a piece of stone
  An elderly naval man.

His hair was weedy, his beard was long,
  And weedy and long was he,
And I heard this wight on the shore recite,
  In a singular minor key:

'Oh, I am a cook and a captain bold,
  And the mate of the *Nancy* brig,
And a bo'sun tight, and a midshipmite,
  And the crew of the captain's gig.'

And he shook his fists and he tore his hair,
   Till I really felt afraid,
For I couldn't help thinking the man had been drinking,
   And so I simply said:

'Oh, elderly man, it's little I know
   Of the duties of men of the sea,
But I'll eat my hand if I understand
   How you can possibly be

'At once a cook, and a captain bold,
   And the mate of the *Nancy* brig,
And a bo'sun tight, and a midshipmite,
   And the crew of the captain's gig.'

Then he gave a hitch to his trousers, which
   Is a trick all seamen larn,
And having got rid of a thumping quid,
   He spun this painful yarn:

' 'Twas in the good ship *Nancy Bell*
   That we sailed to the Indian sea,
And there on a reef we come to grief,
   Which has often occurred to me.

'And pretty nigh all o' the crew was drowned
   (There was seventy-seven o' soul),
And only ten of the *Nancy's* men
   Said "Here!" to the muster-roll.

'There was me and the cook and the captain bold,
   And the mate of the *Nancy* brig,
And the bo'sun tight, and a midshipmite,
   And the crew of the captain's gig.

'For a month we'd neither wittles nor drink,
   Till a-hungry we did feel,
So we drawed a lot, and accordin' shot
   The captain for our meal.

'The next lot fell to the *Nancy's* mate,
And a delicate dish he made;
Then our appetite with the midshipmite
We seven survivors stayed.

'And then we murdered the bo'sun tight,
And he much resembled pig;
Then we wittled free, did the cook and me,
On the crew of the captain's gig.

'Then only the cook and me was left,
And the delicate question, "Which
Of us two goes to the kettle?" arose
And we argued it out as sich.

'For I loved that cook as a brother, I did,
And the cook he worshipped me;
But we'd both be blowed if we'd either be stowed
In the other chap's hold, you see.

' "I'll be eat if you dines off me," says TOM,
"Yes, that," says I, "you'll be,"—
"I'm boiled if I die, my friend," quoth I,
And "Exactly so," quoth he.

'Says he, "Dear JAMES, to murder me
Were a foolish thing to do,
For don't you see that you can't cook *me*,
While I can—and will—cook *you!*"

'So he boils the water, and takes the salt
And the pepper in portions true
(Which he never forgot), and some chopped shalot,
And some sage and parsley too.

' "Come here," says he, with a proper pride,
Which his smiling features tell,
" 'Twill soothing be if I let you see,
How extremely nice you'll smell."

'And he stirred it round and round and round,
  And he sniffed at the foaming froth;
When I ups with his heels, and smothers his squeals
  In the scum of the boiling broth.

'And I eat that cook in a week or less,
  And—as I eating be
The last of his chops, why, I almost drops,
  For a wessel in sight I see!

'And I never grin, and I never smile,
  And I never larf nor play,
But I sit and croak, and a single joke
  I have—which is to say:

'Oh, I am a cook and a captain bold,
  And the mate of the *Nancy* brig,
And a bo'sun tight, *and* a midshipmite,
  And the crew of the captain's gig!'

100 *A Nightmare*

WHEN you're lying awake with a dismal headache, and repose is taboo'd by anxiety,
I conceive you may use any language you choose to indulge in without impropriety;
For your brain is on fire—the bedclothes conspire of usual slumber to plunder you:
First your counterpane goes and uncovers your toes, and your sheet slips demurely from under you;
Then the blanketing tickles—you feel like mixed pickles, so terribly sharp is the pricking,
And you're hot, and you're cross, and you tumble and toss till there's nothing 'twixt you and the ticking.
Then the bedclothes all creep to the ground in a heap, and you pick 'em all up in a tangle;
Next your pillow resigns and politely declines to remain at its usual angle!
Well, you get some repose in the form of a doze, with hot eyeballs and head ever aching,

But your slumbering teems with such horrible dreams that you'd very much better be waking;
For you dream you are crossing the Channel, and tossing about in a steamer from Harwich,
Which is something between a large bathing-machine and a very small second-class carriage;
And you're giving a treat (penny ice and cold meat) to a party of friends and relations—
They're a ravenous horde—and they all came on board at Sloane Square and South Kensington Stations.
And bound on that journey you find your attorney (who started that morning from Devon);
He's a bit undersized, and you don't feel surprised when he tells you he's only eleven.
Well, you're driving like mad with this singular lad (by the bye the ship's now a four-wheeler),
And you're playing round games, and he calls you bad names when you tell him that 'ties pay the dealer';
But this you can't stand, so you throw up your hand, and you find you're as cold as an icicle,
In your shirt and your socks (the black silk with gold clocks), crossing Salisbury Plain on a bicycle:
And he and the crew are on bicycles too—which they've somehow or other invested in—
And he's telling the tars all the particu*lars* of a company he's interested in—
It's a scheme of devices, to get at low prices, all goods from cough mixtures to cables
(Which tickled the sailors) by treating retailers, as though they were all vege*ta*bles—
You get a good spadesman to plant a small tradesman (first take off his boots with a boot-tree),
And his legs will take root, and his fingers will shoot, and they'll blossom and bud like a fruit-tree—
From the greengrocer tree you get grapes and green pea, cauliflower, pineapple, and cranberries,
While the pastry-cook plant cherry-brandy will grant—apple puffs, and three-corners, and banberries—
The shares are a penny, and ever so many are taken by ROTHSCHILD and BARING,
And just as a few are allotted to you, you awake with a shudder despairing—
You're a regular wreck, with a crick in your neck, and no wonder you

snore, for your head's on the floor, and you've needles and pins from your soles to your shins, and your flesh is a-creep, for your left leg's asleep, and you've cramp in your toes, and a fly on your nose, and some fluff in your lung, and a feverish tongue, and a thirst that's intense, and a general sense that you haven't been sleeping in clover;
But the darkness has passed, and it's daylight at last, and and the night has been long—ditto, ditto my song—and thank goodness they're both of them over!

HENRY S. LEIGH

1837–1883

101

Rhymes (?)

My life—to Discontent a prey—
  Is in the sere and yellow leaf.
'Tis vain for happiness to pray:
  No solace brings my heart relief.
My pulse is weak, my spirit low;
  I cannot think, I cannot write.
I strive to spin a verse—but lo!
  My rhymes are very rarely right.

I sit within my lowly cell,
  And strive to court the comic Muse;
But how can Poesy excel,
  With such a row from yonder mews?
In accents passionately high
  The carter chides the stubborn horse;
And shouts a 'Gee!' or yells a 'Hi!'
  In tones objectionably hoarse.

In vain for Poesy I wait;
  No comic Muse my call obeys.
My brains are loaded with a weight
  That mocks the laurels and the bays.
I wish my brains could only be
  Inspired with industry anew;
And labour like the busy bee,
  In strains no Genius ever knew.

Although I strive with all my might,
  Alas, my efforts all are vain!
I've no *afflatus*—not a mite;
  I cannot work the comic vein.
The Tragic Muse may hear my pleas,
  And waft me to a purer clime.
Melpomene! assist me, please,
  To somewhat higher heights to climb.

BRET HARTE

1839–1902

102 *Plain Language from Truthful James*

WHICH I wish to remark,
  And my language is plain,
That for ways that are dark
  And for tricks that are vain,
The heathen Chinee is peculiar,
  Which the same I would rise to explain.

Ah Sin was his name;
  And I shall not deny
In regard to the same,
  What that name might imply;
But his smile it was pensive and childlike,
  As I frequent remarked to Bill Nye.

It was August the third;
  And quite soft was the skies;
Which it might be inferred
  That Ah Sin was likewise;
Yet he played it that day upon William
  And me in a way I despise.

Which we had a small game,
  And Ah Sin took a hand:
It was Euchre. The same
  He did not understand;
But he smiled as he sat by the table,
  With the smile that was childlike and bland.

Yet the cards they were stocked
  In a way that I grieve,
And my feelings were shocked
  At the state of Nye's sleeve:
Which was stuffed full of aces and bowers,
  And the same with intent to deceive.

But the hands that were played
  By that heathen Chinee,
And the points that he made,
  Were quite frightful to see,—
Till at last he put down a right bower,
  Which the same Nye had dealt unto me.

Then I looked up at Nye,
  And he gazed upon me;
And he rose with a sigh,
  And said, 'Can this be?
We are ruined by Chinese cheap labour,—'
  And he went for that heathen Chinee.

In the scene that ensued
  I did not take a hand,
But the floor it was strewed
  Like the leaves on the strand
With the cards that Ah Sin had been hiding,
  In the game 'he did not understand.'

In his sleeves, which were long,
  He had twenty-four packs,—
Which was coming it strong,
  Yet I state but the facts;
And we found on his nails, which were taper,
  What is frequent in tapers,—that's wax.

Which is why I remark,
  And my language is plain,
That for ways that are dark,
  And for tricks that are vain,
The heathen Chinee is peculiar,—
  Which the same I am free to maintain.

103 *Further Language from Truthful James*

Do I sleep? do I dream?
  Do I wonder and doubt?
Are things what they seem?
  Or is visions about?
Is our civilisation a failure?
  Or is the Caucasian played out?

Which expressions are strong;
  Yet would feebly imply
Some account of a wrong—
  Not to call it a lie—
As was worked off on William, my pardner,
  And the same being W. Nye.

He came down to the Ford
  On the very same day
Of that lottery drawed
  By those sharps at the Bay;
And he says to me, 'Truthful, how goes it?'
  I replied, 'It is far, far from gay;

'For the camp has gone wild
  On this lottery game,
And has even beguiled
  "Injin Dick" by the same.'
Which said Nye to me, 'Injins is pizen:
  But what is his number, eh, James?'

I replied, '7,2,
  9,8,4, is his hand';
When he started, and drew
  Out a list, which he scanned;
Then he softly went for his revolver
  With language I cannot command.

Then I said, 'William Nye!'
  But he turned upon me,
And the look in his eye
  Was quite painful to see;
And he says, 'You mistake; this poor Injin
  I protects from such sharps as *you* be!'

I was shocked and withdrew;
  But I grieve to relate,
When he next met my view
  Injin Dick was his mate;
And the two around town was a-lying
  In a frightfully dissolute state.

Which the war dance they had
  Round a tree at the Bend
Was a sight that was sad;
  And it seemed that the end
Would not justify the proceedings,
  As I quietly remarked to a friend.

For that Injin he fled
  The next day to his band;
And we found William spread
  Very loose on the strand,
With a peaceful-like smile on his features.
  And a dollar greenback in his hand;

Which the same, when rolled out,
  We observed with surprise,
Was what he, no doubt,
  Thought the number and prize—
Them figures in red in the corner,
  Which the number of notes specifies.

Was it guile, or a dream?
  Is it Nye that I doubt?
Are things what they seem?
  Or is visions about?
Is our civilisation a failure?
  Or is the Caucasian played out?

AUSTIN DOBSON

1840–1921

104

Dora versus Rose

FROM the tragic-est novels at Mudie's—
  At least, on a practical plan—
To the tales of mere Hodges and Judys,
  One love is enough for a man.
But no case that I ever yet met is
  Like mine: I am equally fond
Of Rose, who a charming brunette is,
        And Dora, a blonde.

Each rivals the other in powers—
  Each waltzes, each warbles, each paints—
Miss Rose, chiefly tumble-down towers;
  Miss Do., perpendicular saints.
In short, to distinguish is folly;
  'Twixt the pair I am come to the pass
Of Macheath, between Lucy and Polly,—
        Or Buridan's ass.

If it happens that Rosa I've singled
  For a soft celebration in rhyme,
Then the ringlets of Dora get mingled
  Somehow with the tune and the time;
Or I painfully pen me a sonnet
  To an eyebrow intended for Do.'s,
And behold I am writing upon it
        The legend, 'To Rose.'

Or I try to draw Dora (my blotter
  Is all overscrawled with her head,)
If I fancy at last that I've got her,
  It turns to her rival instead;
Or I find myself placidly adding
  To the rapturous tresses of Rose
Miss Dora's bud-mouth, and her madding,
        Ineffable nose.

Was there ever so sad a dilemma?
  For Rose I would perish (*pro tem.*);
For Dora I'd willingly stem a—
  (Whatever might offer to stem);
But to make the invidious election,—
  To declare that on either one's side
I've a scruple,—a grain, more affection,
  I *cannot* decide.

And, as either so hopelessly nice is,
  My sole and final resource
Is to wait some indefinite crisis,—
  Some feat of molecular force,
To solve me this riddle conducive
  By no means to peace or repose,
Since the issue can scarce be inclusive
  Of Dora *and* Rose.

(*Afterthought*)
But, perhaps, if a third (say a Norah),
  Not quite so delightful as Rose,—
Not wholly so charming as Dora,—
  Should appear, is it wrong to suppose,—
As the claims of the others are equal,—
  And flight—in the main—is the best,—
That I might . . . But no matter, the sequel
  Is easily guessed.

THOMAS HARDY

1840–1928

105 *The Levelled Churchyard*

'O PASSENGER, pray list and catch
  Our sighs and piteous groans,
Half stifled in this jumbled patch
  Of wrenched memorial stones!

'We late-lamented, resting here,
  Are mixed to human jam,
And each to each exclaims in fear,
  "I know not which I am!"

'The wicked people have annexed
  The verses on the good;
A roaring drunkard sports the text
  Teetotal Tommy should!

'Where we are huddled none can trace,
  And if our names remain,
They pave some path or porch or place
  Where we have never lain!

'Here's not a modest maiden elf
  But dreads the final Trumpet,
Lest half of her should rise herself,
  And half some sturdy strumpet!

'From restorations of Thy fane,
  From smoothings of Thy sward,
From zealous Churchmen's pick and plane
  Deliver us O Lord! Amen!'

106 *The Ruined Maid*

'O 'MELIA, my dear, this does everything crown!
Who could have supposed I should meet you in Town?
And whence such fair garments, such prosperi-ty?'—
'O didn't you know I'd been ruined?' said she.

—'You left us in tatters, without shoes or socks,
Tired of digging potatoes, and spudding up docks;
And now you've gay bracelets and bright feathers three!'—
'Yes: that's how we dress when we're ruined,' said she.

—'At home in the barton you said "thee" and "thou",
And "thik oon", and "theäs oon", and "t'other"; but now
Your talking quite fits 'ee for high compa-ny!'—
'Some polish is gained with one's ruin,' said she.

—'Your hands were like paws then, your face blue and bleak
But now I'm bewitched by your delicate cheek,
And your little gloves fit as on any la-dy!'—
'We never do work when we're ruined,' said she.

—'You used to call home-life a hag-ridden dream,
And you'd sigh, and you'd sock; but at present you seem
To know not of megrims or melancho-ly!'—
'True. One's pretty lively when ruined,' said she.

—'I wish I had feathers, a fine sweeping gown,
And a delicate face, and could strut about Town!'—
'My dear—a raw country girl, such as you be,
Cannot quite expect that. You ain't ruined,' said she.

E. H. PALMER

1840–1882

107

The Parterre

I DON'T know any greatest treat
As sit him in a gay parterre,
And sniff one up the perfume sweet
Of every roses buttoning there.

It only want my charming miss
Who make to blush the self red rose;
Oh! I have envy of to kiss
The end's tip of her splendid nose.

Oh! I have envy of to be
What grass 'neath her pantoffle push,
And too much happy seemeth me
The margaret which her vestige crush.

But I will meet her nose at nose,
And take occasion for her hairs,
And indicate her all my woes,
That she in fine agree my prayers.

I don't know any greatest treat
As sit him in a gay parterre,
With Madame who is too more sweet
Than every roses buttoning there.

ANDREW LANG

1844–1912

108

Brahma

IF the wild bowler thinks he bowls,
  Or if the batsman thinks he's bowled,
They know not, poor misguided souls,
  They too shall perish unconsoled.
I am the batsman and the bat,
  I am the bowler and the ball,
The umpire, the pavilion cat,
  The roller, pitch, and stumps, and all.

GERARD MANLEY HOPKINS

1844–1889

109

Triolet

'THE child is father to the man.'
How can he be? The words are wild.
Suck any sense from that who can:
'The child is father to the man.'
No; what the poet did write ran,
'The man is father to the child.'
'The child is father to the man'!
How *can* he be? The words are wild.

R. L. STEVENSON

1850–1894

110

Not I

Some like drink
  In a pint pot,
Some like to think;
  Some not.

Strong Dutch cheese,
  Old Kentucky Rye,
Some like these;
  Not I.

Some like Poe,
  And others like Scott,
Some like Mrs. Stowe;
  Some not.

Some like to laugh,
  Some like to cry,
Some like chaff;
  Not I.

ANONYMOUS

111

She was Poor but she was Honest

She was poor but she was honest,
  Victim of a rich man's game;
First he loved her, then he left her,
  And she lost her maiden name.

Then she hastened up to London,
  For to hide her grief and shame;
There she met another rich man,
  And she lost her name again.

See her riding in her carriage,
  In the Park and all so gay;
All the nibs and nobby persons
  Come to pass the time of day.

See them at the gay theáter
  Sitting in the costly stalls;
With one hand she holds the programme,
  With the other strokes his hand.

See him have her dance in Paris
  In her frilly underclothes;
All those Frenchies there applauding
  When she strikes a striking pose.

See the little country village
  Where her aged parents live;
Though they drink champagne she sends them,
  Still they never can forgive.

In the rich man's arms she flutters
  Like a bird with a broken wing;
First he loved her, then he left her,
  And she hasn't got a ring.

See him in his splendid mansion
  Entertaining with the best,
While the girl as he has ruined
  Entertains a sordid guest.

See him riding in his carriage
  Past the gutter where she stands;
He has made a stylish marriage
  While she wrings her ringless hands.

See him in the House of Commons
  Passing laws to put down crime,
While the victim of his passions
  Slinks away to hide her shame.

See her on the bridge at midnight
  Crying, 'Farewell, faithless love!'
There's a scream, a splash—Good Heavens!
  What is she a-doing of?

Then they dragged her from the river,
  Water from her clothes they wrung;
They all thought that she was drownded,
  But the corpse got up and sung:

'It's the same the whole world over;
  It's the poor as gets the blame,
It's the rich as gets the pleasure—
  Ain't it all a bleeding shame!'

A. C. HILTON

1851–1877

112 *The Heathen Pass-ee*

WHICH I wish to remark,
  And my language is plain,
That for plots that are dark
  And not always in vain,
The Heathen Pass-ee is peculiar,
  And the same I would rise to explain.

I would also premise
  That the term of Pass-ee
Most fitly applies,
  As you probably see,
To one whose vocation is passing
  The 'ordinary B.A. degree.'

A. C. HILTON

Tom Crib was his name,
  And I shall not deny
In regard to the same
  What that name might imply,
That his face it was trustful and childlike,
  And he had the most innocent eye.

Upon April the First
  The Little-Go fell,
And that was the worst
  Of the gentleman's sell,
For he fooled the Examining Body
  In a way I'm reluctant to tell.

The candidates came
  And Tom Crib soon appeared;
It was Euclid, the same
  Was 'the subject he feared';
But he smiled as he sat by the table
  With a smile that was wary and weird.

Yet he did what he could,
  And the papers he showed
Were remarkably good,
  And his countenance glowed
With pride when I met him soon after
  As he walked down the Trumpington Road.

We did not find him out,
  Which I bitterly grieve,
For I've not the least doubt
  That he'd placed up his sleeve
Mr. Todbunker's excellent Euclid,
  The same with intent to deceive.

But I shall not forget
  How the next day or two
A stiff paper was set
  By Examiner U——
On Euripides' tragedy, Bacchae,
  A subject Tom 'partially knew.'

But the knowledge displayed
  By that Heathen Pass-ee,
And the answers he made
  Were quite frightful to see,
For he rapidly floored the whole paper
  By about twenty minutes to three.

Then I looked up at U——
  And he gazed upon me,
I observed, 'This won't do';
  He replied, 'Goodness me!
We are fooled by this artful young person.'
  And he sent for that Heathen Pass-ee.

The scene that ensued
  Was disgraceful to view,
For the floor it was strewed
  With a tolerable few
Of the 'tips' that Tom Crib had been hiding
  For the 'subject he partially knew.'

On the cuff of his shirt
  He had managed to get
What we hoped had been dirt,
  But which proved, I regret,
To be notes on the rise of the Drama,
  A question invariably set.

In his various coats
  We proceeded to seek,
Where we found sundry notes
  And—with sorrow I speak—
One of Bohn's publications, so useful
  To the student of Latin or Greek.

In the crown of his cap
  Were the Furies and Fates,
And a delicate map
  Of the Dorian States,
And we found in his palms, which were hollow,
  What are frequent in palms—that is, dates;

Which is why I remark,
  And my language is plain,
That for plots that are dark
  And not always in vain,
The Heathen Pass-ee is familiar,
  Which the same I am free to maintain.

A. D. GODLEY

1856–1925

113

After Horace

WHAT asks the Bard? He prays for naught
  But what the truly virtuous crave:
That is, the things he plainly ought
    To have.

'Tis not for wealth, with all the shocks
  That vex distracted millionaires,
Plagued by their fluctuating stocks
    And shares:

While plutocrats their millions new
  Expend upon each costly whim,
A great deal less than theirs will do
    For him:

The simple incomes of the poor
  His meek poetic soul content:
Say, £30,000 at four
    Per cent.!

His taste in residence is plain:
  No palaces his heart rejoice:
A cottage in a lane (Park Lane
    For choice)—

Here be his days in quiet spent:
  Here let him meditate the Muse:
Baronial Halls were only meant
          For Jews,

And lands that stretch with endless span
  From east to west, from south to north,
Are often much more trouble than
          They're worth!

Let epicures who eat too much
  Become uncomfortably stout:
Let gourmets feel th' approaching touch
          Of gout,—

The Bard subsists on simpler food:
  A dinner, not severely plain,
A pint or so of really good
          Champagne—

Grant him but these, no care he'll take
  Though Laureates bask in Fortune's smile,
Though Kiplings and Corellis make
          Their pile:

Contented with a scantier dole
  His humble Muse serenely jogs,
Remote from scenes where authors roll
          Their logs:

Far from the madding crowd she lurks,
  And really cares no single jot
Whether the public read her works
          Or not!

114 *The Motor Bus*

WHAT is this that roareth thus?
Can it be a Motor Bus?
Yes, the smell and hideous hum
Indicant Motorem Bum!
Implet in the Corn and High
Terror me Motoris Bi:
Bo Motori clamitabo
Ne Motore caedar a Bo—
Dative be or Ablative
So thou only let us live:—
Whither shall thy victims flee?
Spare us, spare us, Motor Be!
Thus I sang; and still anigh
Came in hordes Motores Bi,
Et complebat omne forum
Copia Motorum Borum.
How shall wretches live like us
Cincti Bis Motoribus?
Domine, defende nos
Contra hos Motores Bos!

115 *Women's Degrees*

A TANGLED web indeed we weave
When Adam grants degrees to Eve:
And much I doubt, had Eve first had 'em,
If she'd have done as much for Adam.

J. K. STEPHEN

1859–1892

116

On a Parisian Boulevard

BRITANNIA rules the waves,
As I have heard her say;
She frees whatever slaves
She meets upon her way.

A teeming mother she
Of Parliaments and Laws;
Majestic, mighty, free:
Devoid of common flaws.

For her did Shakespere write
His admirable plays;
For her did Nelson fight
And Wolseley win his bays.

Her sturdy common sense
Is based on solid grounds:
By saving numerous pence
She spends effective pounds.

The Saxon and the Celt
She equitably rules;
Her iron rod is felt
By countless knaves and fools.

In fact, mankind at large,
Black, yellow, white and red,
Is given to her in charge
And owns her as a head.

But every here and there—
Deny it if you can—
She breeds a vacant stare
Unworthy of a man:

A look of dull surprise;
  A nerveless idle hand:
An eye which never tries
  To threaten or command:

In short, a kind of man,
  If man indeed he be,
As worthy of our ban
  As any that we see:

Unspeakably obtuse,
  Abominably vain,
Of very little use,
  And execrably plain.

117 *On a Rhine Steamer*

REPUBLIC of the West
  Enlightened, free, sublime,
Unquestionably best
  Production of our time,

The telephone is thine,
  And thine the Pullman car,
The caucus, the divine
  Intense electric star.

To thee we likewise owe
  The venerable names
Of Edgar Allan Poe,
  And Mr. Henry James.

In short it's due to thee,
  Thou kind of Western star,
That we have come to be
  Precisely what we are.

But every now and then,
  It cannot be denied,
You breed a kind of men
  Who are not dignified,

Or courteous or refined,
  Benevolent or wise,
Or gifted with a mind
  Beyond the common size,

Or notable for tact,
  Agreeable to me,
Or anything, in fact,
  That people ought to be.

118

Drinking Song

THERE are people, I know, to be found,
  Who say, and apparently think,
That sorrow and care may be drowned
  By a timely consumption of drink.

Does not man, these enthusiasts ask,
  Most nearly approach the divine,
When engaged in the soul-stirring task
  Of filling his body with wine?

Have not beggars been frequently known,
  When satisfied, soaked and replete,
To imagine their bench was a throne
  And the civilised world at their feet?

Lord Byron has finely described
  The remarkably soothing effect
Of liquor, profusely imbibed,
  On a soul that is shattered and wrecked.

In short, if your body or mind
  Or your soul or your purse come to grief,
You need only get drunk, and you'll find
  Complete and immediate relief.

For myself, I have managed to do
  Without having recourse to this plan,
So I can't write a poem for you,
  And you'd better get someone who can.

119

Sincere Flattery

(i)

(*Of R.B.*)

BIRTHDAYS? yes, in a general way;
For the most if not for the best of men:
You were born (I suppose) on a certain day:
So was I: or perhaps in the night: what then?

Only this: or at least, if more,
You must know, not think it, and learn, not speak:
There is truth to be found on the unknown shore,
And many will find where few will seek.

For many are called and few are chosen,
And the few grow many as ages lapse:
But when will the many grow few: what dozen
Is fused into one by Time's hammer-taps?

A bare brown stone in a babbling brook:—
It was wanton to hurl it there, you say:
And the moss, which clung in the sheltered nook
(Yet the stream runs cooler), is washed away.

That begs the question: many a prater
Thinks such a suggestion a sound 'stop thief!'
Which, may I ask, do you think the greater,
Sergeant-at-arms or a Robber Chief?

And if it were not so? still you doubt?
Ah! yours is a birthday indeed if so.
That were something to write a poem about,
If one thought a little. I only know.

P.S.

There's a Me Society down at Cambridge,
Where my works, *cum notis variorum*,
Are talked about; well, I require the same bridge
That Euclid took toll at as *Asinorum*:

And, as they have got through several ditties
I thought were as stiff as a brick-built wall,
I've composed the above, and a stiff one *it* is,
A bridge to stop asses at, once for all.

(ii)
(*Of W.W.*)

THE clear cool note of the cuckoo which has ousted the legitimate nest-holder,
The whistle of the railway guard dispatching the train to the inevitable collision,
The maiden's monosyllabic reply to a polysyllabic proposal,
The fundamental note of the last trump, which is presumably D natural;
All of these are sounds to rejoice in, yea, to let your very ribs re-echo with:
But better than all of them is the absolutely last chord of the apparently inexhaustible pianoforte player.

120 *To R.K.*

As long I dwell on some stupendous
And tremendous (Heaven defend us!)
Monstr'-inform'-ingens-horrendous
Demoniaco-seraphic
Penman's latest piece of graphic.

BROWNING.

WILL there never come a season
Which shall rid us from the curse
Of a prose which knows no reason
And an unmelodious verse:
When the world shall cease to wonder
At the genius of an Ass,
And a boy's eccentric blunder
Shall not bring success to pass:

When mankind shall be delivered
From the clash of magazines,
And the inkstand shall be shivered
Into countless smithereens:
When there stands a muzzled stripling,
Mute, beside a muzzled bore:
When the Rudyards cease from Kipling
And the Haggards Ride no more.

121

A Sonnet

Two voices are there: one is of the deep;
It learns the storm-cloud's thunderous melody,
Now roars, now murmurs with the changing sea,
Now bird-like pipes, now closes soft in sleep:
And one is of an old half-witted sheep
Which bleats articulate monotony,
And indicates that two and one are three,
That grass is green, lakes damp, and mountains steep:
And, Wordsworth, both are thine: at certain times
Forth from the heart of thy melodious rhymes,
The form and pressure of high thoughts will burst:
At other times—good Lord! I'd rather be
Quite unacquainted with the ABC
Than write such hopeless rubbish as thy worst.

A. E. HOUSMAN

1859–1936

122

Fragment of a Greek Tragedy

Alcmaeon, Chorus

CHO. O SUITABLY-attired-in-leather-boots
Head of a traveller, wherefore seeking whom
Whence by what way how purposed art thou come
To this well-nightingaled vicinity?
My object in enquiring is to know,

But if you happen to be deaf and dumb
And do not understand a word I say,
Then wave your hand, to signify as much.

ALC. I journeyed hither a Bœotain road.
CHO. Sailing on horseback, or with feet for oars?
ALC. Plying with speed my partnership of legs.
CHO. Beneath a shining or a rainy Zeus?
ALC. Mud's sister, not himself, adorns my shoes.
CHO. To learn your name would not displease me much.
ALC. Not all that men desire do they obtain.
CHO. Might I then hear at what your presence shoots?
ALC. A shepherd's questioned mouth informed me that—
CHO. What? for I know not yet what you will say—
ALC. Nor will you ever, if you interrupt.
CHO Proceed, and I will hold my speechless tongue.
ALC. —This house was Eriphyla's, no one's else.
CHO. Nor did he shame his throat with hateful lies.
ALC. May I then enter, passing through the door?
CHO. Go, chase into the house a lucky foot,
And, O my son, be, on the one hand, good,
And do not, on the other hand, be bad;
For that is very much the safest plan.
ALC. I go into the house with heels and speed.

Chorus

In speculation *Strophe*
I would not willingly acquire a name
For ill-digested thought;
But after pondering much
To this conclusion I at last have come:
Life is uncertain.
This truth I have written deep
In my reflective midriff
On tablets not of wax,
Nor with a pen did I inscribe it there,
For many reasons: *Life, I say, is not*
A stranger to uncertainty.
Not from the flight of omen-yelling fowls
This fact did I discover.
Nor did the Delphic tripod bark it out,
Nor yet Dodona.
Its native ingenuity sufficed
My self-taught diaphragm.

Antistrophe

Why should I mention
The Inachean daughter, loved of Zeus?
Her whom of old the gods,
More provident than kind,
Provided with four hoofs, two horns, one tail,
A gift not asked for,
And sent her forth to learn
The unfamiliar science
Of how to chew the cud.
She therefore, all about the Argive fields,
Went cropping pale green grass and nettle-tops,
Nor did they disagree with her.
But yet, howe'er nutritious, such repasts
I do not hanker after:
Never may Cypris for her seat select
My dappled liver!
Why should I mention Io? Why indeed?
I have no notion why.

Epode

But now does my boding heart,
Unhired, unaccompanied, sing
A strain not meet for the dance.
Yea even the palace appears
To my yoke of circular eyes
(The right, nor omit I the left)
Like a slaughterhouse, so to speak,
Garnished with woolly deaths
And many shipwrecks of cows.
I therefore in a Cissian strain lament;
And to the rapid,
Loud, linen-tattering thumps upon my chest
Resounds in concert
The battering of my unlucky head.

ERIPHYLA (*within*). O, I am smitten with a hatchet's jaw;
And that in deed and not in word alone.

CHO. I thought I heard a sound within the house
Unlike the voice of one that jumps for joy.

ERI. He splits my skull, not in a friendly way,
One more: he purposes to kill me dead.

CHO. I would not be reputed rash, but yet
I doubt if all be gay within the house.

ERI. O! O! another stroke! that makes the third.
He stabs me to the heart against my wish.
CHO. If that be so, thy state of health is poor;
But thine arithmetic is quite correct.

123

Occasional Poem

WHEN Adam day by day
Woke up in Paradise,
He always used to say
'Oh, this is very nice.'

But Eve from scenes of bliss
Transported him for life.
The more I think of this
The more I beat my wife.

124

The Elephant, or The Force of Habit

A TAIL behind, a trunk in front,
Complete the usual elephant.
The tail in front, the trunk behind
Is what you very seldom find.

If you for specimens should hunt
With trunks behind and tails in front,
That hunt would occupy you long;
The force of habit is so strong.

125

Infant Innocence

READER, behold! this monster wild
Has gobbled up the infant child.
The infant child is not aware
It has been eaten by the bear.

KENNETH GRAHAME

1859–1932

126

The Song of Mr. Toad

THE world has held great Heroes,
 As history books have showed;
But never a name to go down to fame
 Compared with that of Toad!

The clever men at Oxford
 Know all that there is to be knowed,
But they none of them knew one half as much
 As intelligent Mr. Toad!

The animals sat in the Ark and cried,
 Their tears in torrents flowed.
Who was it said, 'There's land ahead'?
 Encouraging Mr. Toad!

The Army all saluted
 As they marched along the road.
Was it the King? Or Kitchener?
 No. It was Mr. Toad!

The Queen and her Ladies-in-waiting
 Sat at the window and sewed.
She cried, 'Look! who's that *handsome* man?'
 They answered, 'Mr. Toad.'

127

BALLIOL RHYMES

(i)

FIRST come I; my name is Jowett.
There's no knowledge but I know it.
I am Master of this college:
What I don't know isn't knowledge.

H. C. BEECHING (1859–1919)

(ii)

I AM the Dean of Christ Church, Sir:
There's my wife; look well at her.
She's the Broad and I'm the High;
We are the University.

C. A. SPRING-RICE (1858–1918)

(iii)

MY name is George Nathaniel Curzon,
I am a most superior person.
My face is pink, my hair is sleek,
I dine at Blenheim once a week.

ANON.

(iv)

I AM tall and rather stately,
And I care not very greatly
What you say, or what you do.
I'm Mackail,—and who are you?

ANON.

ANONYMOUS

128

Gasbags

I'M thankful that the sun and moon
Are both hung up so high
That no pretentious hand can stretch
And pull them from the sky.
If they were not, I have no doubt
But some reforming ass
Would recommend to take them down
And light the world with gas.

129 *The Tale of Lord Lovell*

LORD LOVELL he stood at his own front door,
  Seeking the hole for the key;
His hat was wrecked, and his trousers bore
  A rent across either knee,
When down came the beauteous Lady Jane
  In fair white draperie.

'Oh, where have you been, Lord Lovell?' she said,
  'Oh, where have you been?' said she;
'I have not closed an eye in bed,
  And the clock has just struck three.
Who has been standing you on your head
  In the ash-barrel, pardie?'

'I am not drunk, Lad' Shane,' he said:
  'And so late it cannot be;
The clock struck one as I enterèd—
  I heard it two times or three;
It must be the salmon on which I fed
  Has been too many for me.'

'Go tell your tale, Lord Lovell,' she said,
  'To the maritime cavalree,
To your grandmother of the hoary head—
  To any one but me:
The door is not used to be openèd
  With a cigarette for a key.'

130 *The Ballad of William Bloat*

IN a mean abode in the Shankill Road
  Lived a man named William Bloat;
Now he had a wife, the plague of his life,
  Who continually got his goat,
And one day at dawn, with her night-shift on,
  He slit her bloody throat.

With a razor-gash he settled her hash—
  Oh, never was death so quick;
But the steady drop on the pillowslip
  Of her life-blood turned him sick,
And the pool of gore on the bedroom floor
  Grew clotted and cold and thick.

Now he was right glad he had done as he had
  As his wife lay there so still,
When a sudden awe of the mighty Law
  Struck his heart with an icy chill,
And, to finish the fun so well begun,
  He resolved himself to kill.

He took the sheet from his wife's cold feet
  And knotted it into a rope,
And hanged himself from the pantry shelf—
  An easy death, let's hope.
In the jaws of death with his latest breath
  Said he, 'To Hell with the Pope'.

But the strangest turn of the whole concern
  Is only just beginning:
He went to Hell, but his wife got well
  And is still alive and sinning,
For the razor-blade was Dublin-made
  But the sheet was Belfast linen.

WALTER RALEIGH

1861–1922

131

Wishes of an Elderly Man

I WISH I loved the Human Race;
I wish I loved its silly face;
I wish I liked the way it walks;
I wish I liked the way it talks;
And when I'm introduced to one
I wish I thought *What Jolly Fun*!

OWEN SEAMAN

1861–1936

132

A Birthday Ode to Mr. Alfred Austin

THE early bird got up and whet his beak;
  The early worm arose, an easy prey;
This happened any morning in the week,
    Much as today.

The moke uplift for joy his hinder hoof;
  Shivered the fancy poodle, freshly shorn;
The prodigal upon the attic roof
    Mewed to the morn.

His virile note the cock profusely blew;
  The beetle trotted down the kitchen tong;
The early bird above alluded to
    Was going strong.

All this of course refers to England's isle,
  But things were going on across the deep;
In Egypt—take a case—the crocodile
    Was sound asleep.

Buzzed the Hymettian bee; sat up in bed
  The foreign oyster sipping local drains;
The impious cassowary lay like lead
    On Afric's plains.

A-nutting went the nimble chimpanzee;—
  And what, you ask me, am I driving at?
Wait on: in less than twenty minutes we
    Shall come to that.

The bulbous crowfoot drained his dewy cup;
  The saxifrage enjoyed a morning crawl;
The ampelopsis slowly sidled up
    The garden wall.

Her petals wide the periwinkle flung;
  Blue gentian winked upon unweaned lambs;
And there was quite a pleasant stir among
        The cryptogams.

May was the month alike in croft and wild
  When—here, in fact, begins the actual tale—
When forth withal there came an infant child,
        A healthy male.

Marred was his ruby countenance, as when
  A blushing peony is moist with rain;
And first he strenuously kicked, and then
        He kicked again.

They put the bays upon his barren crest,
  Laid on his lap a lexicon of rhyme,
Saying—'You shall with luck attain the quest,
        In course of time.'

Stolid he gazed, as one that may not know
  The meaning of a presage—or is bored;
But when he loosed his lips it was as though
        The sea that roared.

That dreadful summons to a higher place
  He would not, if he could, have spurned away;
But, being a babe, he had, in any case,
        Nothing to say.

*

Eight happy summers passed and Southey too,
  And one that had the pull in point of age
Walked in; for Alfred still was struggling through
        The grammar-stage.

When William followed out in Robert's wake,
  An alien Alfred filled the vacant spot,
Possibly by some clerical mistake,
        Possibly not.

Our friend had then achieved but fifteen years,
Nor yet against him was there aught to quote;
For he had uttered in the nation's ears
Not half a note.

*

At forty-one he let his feelings go:—
'If he, that other Alfred, ever die,
And I am not appointed, I will know
The reason why!'

Some sixteen further autumns bound their sheaves;
With hope deferred wild battle he had waged,
And written books. At last the laurel-leaves
Were disengaged.

Felicitations, bursting through his bowers,
Came on him hoeing roots. With mild surprise,
'Leave me alone,' he said, 'among my flowers
To botanize.'

The Prime Elector, Man of Many Days,
Though Allan's Muse adorned the Liberal side,
Seizing the swift occasion, left the bays
Unoccupied.

The Peer that followed, having some regard
For humour hitherto accounted sin,
Produced a knighthood for the blameless bard
Of proud Penbryn.

At length a callous Tory Chief arose,
Master of caustic jest and cynic gibe,
Looked round the Carlton Club and lightly chose
Its leading scribe.

And so with heaving heart and happy tears
Our patient Alfred took the tardy spoil,
Though spent with sixty venerable years
Of virtuous toil.

And ever, when marsh-marigolds are cheap,
  And new potatoes crown the death of May,
If memory serve us, we propose to keep
      His natal day.

133 *England Expects?*

IF earthward you could wing your flight
  And look on London's central zone,
Seizing that eligible site
  Where stands your counterfeit in stone,
I wonder, Nelson, if your eye
  Would even form the faintest image
Of what emotions underlie
  This tumult, this stupendous scrimmage.

Could you desert that heavenly place
  Where sailors know their pilot-star
To view the many-peopled space
  Named by the name of Trafalgar,
Remembering how your signal ran,
  That still remains a thing of beauty,
You might expect that every man
  This day, as then, would do his duty.

Alas! we have no ships afloat
  Upon the basins in the Square;
It is the landsman's lusty throat
  That rends today a saltless air;
And, save from such as hold the main
  To guard her pride among the nations,
England has ceased to entertain
  Much in the way of expectations.

O yes, they'll shout all right enough!
  It costs them little; noise is cheap;
But have they hearts of quite the stuff
  That made your loyal pulses leap?
They'll roar you till their midriffs ache
  Under the bunting's brave devices,
But wouldn't lift a hand to make
  The least of all your sacrifices.

A wind of words—and nothing more!
  But if the test were sought in deeds,
If England asked the sons she bore
  Each man to serve the Mother's needs,
If she 'expected' such a debt
  To stir the blood of those that owe it,
The sole response she's like to get
  Would be, 'No thanks; not if we know it.'

Just now they pipe a patriot tune;
  Anon they'll wonder why they spent
A precious football afternoon
  Mafficking round a monument;
And myriads who go mad today—
  Give them a week, they'll go yet madder,
Watching the modern heroes' fray,
  Where hirelings hoof a bounding bladder.

Much you would have to marvel at
  Could you return this autumn-tide;
You'd find the Fleet—thank God for that—
  Staunch and alert as when you died;
But, elsewhere, few to play your part,
  Ready at need and ripe for action;
The rest—in idle ease of heart
  Smiling an unctuous satisfaction.

I doubt if you could well endure
  These new ideals (so changed we are)
Undreamed, Horatio, in your
  Philosophy of Trafalgar;
And, should you still 'expect' to see
  The standard reached which you erected,
Nothing just now would seem to be
  So certain as the unexpected.

134 *A Plea for Trigamy*

I'VE been trying to fashion a wifely ideal,
  And find that my tastes are so far from concise
That, to marry completely, no fewer than three'll
  Suffice.

I've subjected my views to severe atmospheric
  Compression, but still, in defiance of force,
They distinctly fall under three heads, like a cleric
  Discourse.

My *first* must be fashion's own fancy-bred daughter,
  Proud, peerless, and perfect—in fact, *comme il faut*;
A waltzer and wit of the very first water—
  For *show*.

But these beauties that serve to make all the men jealous,
  Once face them alone in the family cot,
Heaven's angels incarnate (the novelists tell us)
  They're *not*.

But so much for appearances. Now for my *second*,
  My lover, the wife of my home and my heart:
Of all fortune and fate of my life to be reckon'd
  A part.

She must know all the needs of a rational being,
  Be skilled to keep council, to comfort, to coax;
And, above all things else, be accomplished at seeing
  My jokes.

I complete the ménage by including one other
  With all the domestic prestige of a hen:
As my housekeeper, nurse, or it may be, a mother
  Of men.

Total *three!* and the virtues all well represented;
  With fewer than this such a thing can't be done;
Though I've known married men who declare
  they're contented
  With one.

Would you hunt during harvest, or hay-make in
  winter?
  And how can one woman expect to combine
Certain qualifications essentially inter-
  necine?

You may say that my prospects are (legally) sunless;
  I state that I find them as clear as can be:—
I will marry *no* wife, since I can't do with one less
  Than three.

135 *The Sitting Bard*

FELLOW, you have no *flair* for art, I fear,
  Who thus confound me with the idle Many—
    The loafer pensive o'er his betting rag,
    The messenger (express) with reeking fag,
The nursemaid sighing for her bombardier—
  All charged the same pew-rate, a common penny.[1]

[1] The original charge for a chair in St. James's Park

I am an artist; I am not as these;
He does me horrid despite who confuses
My taste with theirs who come this way to chuck
Light provender to some exotic duck,
Whereas I sit beneath these secular trees
In close collaboration with the Muses.

To me St. James's Park is holy ground;
In fancy I regard these glades as Helicon's;
This lake (although an artificial pond)
To Hippocrene should roughly correspond;
Others, not I, shall make its shores resound,
Bandying chaff with yonder jaunty pelicans.

All this escaped you, lacking minstrel lore.
'Tis so with poets: men are blind and miss us;
You did not mark my eye's exultant mood,
The inflated chest, the listening attitude,
Nor, bent above the mere, the look I wore
When lost in self-reflection—like Narcissus.

Else you could scarce have charged me for my seat;
I must have earned an honorary session;
For how could I have strained your solid chair,
I that am all pure spirit, fine as air,
And sit as light as when with wingèd feet
Mercury settles, leaving no impression?

Well, take your paltry penny, trivial dun!
And bid your chair-contractors freely wallow
In luxury therewith; but, when you find
Another in this hallowed seat reclined,
Squeeze him for tuppence, saying, '*Here sat one*
On June the fifth and parleyed with Apollo.'

RUDYARD KIPLING

1865–1936

136

The Lie

THERE is pleasure in the wet, wet clay,
When the artist's hand is potting it.
There is pleasure in the wet, wet lay,
When the poet's pad is blotting it.
There is pleasure in the shine of your picture
  on the line
At the Royal Acade-my;
But the pleasure felt in these is as chalk to
  Cheddar cheese
When it comes to a well-made Lie.—
To a quite unwreckable Lie,
To a most impeccable Lie!
To a water-tight, fire-proof, angle-iron, sunk-
  hinge, time-lock, steel-faced Lie!
Not a private hansom Lie,
But a pair-and-brougham Lie,
Not a little-place-at-Tooting, but a country-
  house-with-shooting
And a ring-fence-deer-park Lie.

B. L. TAYLOR

1866–1921

137

Canopus

WHEN quacks with pills political would dope us,
  When politics absorbs the livelong day,
I like to think about the star Canopus,
  So far, so far away.

Greatest of visioned suns, they say who list 'em;
  To weigh it science always must despair.
Its shell would hold our whole dinged solar system,
  Nor ever know 'twas there.

When temporary chairmen utter speeches,
  And frenzied henchmen howl their battle hymns,
My thoughts float out across the cosmic reaches
  To where Canopus swims.

When men are calling names and making faces,
  And all the world's ajangle and ajar,
I meditate on interstellar spaces
  And smoke a mild seegar.

For after one has had about a week of
  The arguments of friends as well as foes,
A star that has no parallax to speak of
  Conduces to repose.

HILAIRE BELLOC

1870–1953

138 *On a General Election*

THE accursèd power which stands on Privilege
(And goes with Women, and Champagne and Bridge)
Broke—and Democracy resumed her reign:
(Which goes with Bridge, and Women and Champagne).

139 *Fatigue*

I'M tired of Love: I'm still more tired of Rhyme.
But Money gives me pleasure all the time.

140

Lord Finchley

LORD FINCHLEY tried to mend the Electric Light
Himself. It struck him dead: And serve him right!
It is the business of the wealthy man
To give employment to the artisan.

141

The Yak

AS a friend to the children commend me the Yak,
You will find it exactly the thing:
It will carry and fetch, you can ride on its back,
Or lead it about with a string.

The Tartar who dwells on the plains of Thibet
(A desolate region of snow)
Has for centuries made it a nursery pet,
And surely the Tartar should know!

Then tell your papa where the Yak can be got,
And if he is awfully rich
He will buy you the creature—or else he will *not*.
(I cannot be positive which.)

142

FROM *A Moral Alphabet*

THE Dreadful Dinotherium he
Will have to do his best for D.
The early world observed with awe
His back, indented like a saw.
His look was gay, his voice was strong;
His tail was neither short nor long;
His trunk, or elongated nose,
Was not so large as some suppose;
His teeth, as all the world allows,
Were graminivorous, like a cow's.

He therefore should have wished to pass
Long peaceful nights upon the Grass,
But being mad the brute preferred
To roost in branches, like a bird.[1]
A creature heavier than a whale,
You see at once, could hardly fail
To suffer badly when he slid
And tumbled (as he always did).
His fossil, therefore, comes to light
All broken up: and serve him right.

MORAL

If you were born to walk the ground,
Remain there; do not fool around.

1. We have good reason to suppose
He did so, from his claw-like toes.

143 *Rebecca, who slammed doors for fun and perished miserably*

A TRICK that everyone abhors
In Little Girls is slamming Doors,
A Wealthy Banker's Little Daughter
Who lived in Palace Green, Bayswater
(By name Rebecca Offendort),
Was given to this Furious Sport.

She would deliberately go
And Slam the door like Billy-Ho!
To make her Uncle Jacob start.
She was not really bad at heart,
But only rather rude and wild:
She was an aggravating child . . .

It happened that a Marble Bust
Of Abraham was standing just
Above the Door this little Lamb
Had carefully prepared to Slam,
And Down it came! It knocked her flat!
It laid her out! She looked like that.

Her funeral Sermon (which was long
And followed by a Sacred Song)
Mentioned her Virtues, it is true,
But dwelt upon her Vices too,
And showed the Dreadful End of One
Who goes and slams the door for Fun.

The children who were brought to hear
The awful Tale from far and near
Were much impressed, and inly swore
They never more would slam the Door.
—As often they had done before.

ANTHONY C. DEANE

1870–1946

144

An Ode

I SING a song of sixpence, and of rye
  A pocketful—recalling, sad to state,
The niggardly emoluments which I
  Receive as Laureate!

Also I sing of blackbirds—in the mart
  At four-a-penny. Thus, in other words,
The sixpence which I mentioned at the start
  Purchased two dozen birds.

So four-and-twenty birds were deftly hid—
  Or shall we say, were skilfully concealed?—
Within the pie-dish. When they raised the lid,
  What melody forth pealed!

Now I like four-and-twenty blackbirds sing,
  With all their sweetness, all their rapture keen;
And isn't this a pretty little thing
  To set before the Queen?

The money-counting monarch—sordid man!—
His wife, who robbed the little busy bees,
I disregard. In fact a poet can
But pity folks like these.

The maid was in thc garden. Happy maid!
Her choice entitles her to rank above
Master and Mistress. Gladly she surveyed
The Garden That I Love!

—Where grow my daffodils, anemones,
Tulips, auriculas, chrysanthemums,
Cabbages, asparagus, sweet peas,
With apples, pears, and plums—

(That's a parenthesis. The very name
Of garden really carries one astray!)
But suddenly a feathered ruffian came,
And stole her nose away.

Eight stanzas finished! So my Court costume
I lay aside: the Laureate, I suppose,
Has done his part; the man may now resume
His iournalistic prose.

145 *The Cult of the Celtic*

WHEN the eager squadrons of day are faint and disbanded,
And under the wind-swept stars the reaper gleans
The petulant passion flowers—although, to be candid,
I haven't the faintest notion what that means—

Surely the Snow-White Bird makes melody sweeter
High in the air than skimming the clogging dust.
(Yes, there's certainly something queer about this metre,
But, as it's Celtic, you and I must take it on trust.)

And oh, the smile of the Slave as he shakes his fetters!
And oh, the Purple Pig as it roams afar!
And oh, the—something or other in capital letters—
As it yields to the magic spell of a wind-swept star!

And look at the tricksy Elves, how they leap and frolic,
  Ducking the Bad Banshee in the moonlit pool,
Celtic, yet fully content to be 'symbolic',
  Never a thought in their head about Home Rule!

But the wind-swept star—you notice it has to figure,
  Taking an average merely, in each alternate verse
Of every Celtic poem—smiles with a palpable snigger,
  While the Yellow Wolf-Hound bays his blighting curse,

And the voices of dead desires in sufferers waken,
  And the voice of the limitless lake is harsh and rough,
And the voice of the reader, too, unless I'm mistaken,
  Is heard to remark that he's had about enough.

But since the critics have stated with some decision
  That stanzas very like these are simply grand,
Showing 'a sense of beauty and intimate vision',
  Proving a 'Celtic Renaissance' close at hand;

Then, although I admit it's a terrible tax on
  Powers like mine, yet I sincerely felt
My task, as an unintelligent Saxon,
  Was, at all hazards, to try to copy the Celt!

MAX BEERBOHM

1872–1956

146 *Police Station Ditties*

THEN it's collar 'im tight,
  In the name of the Law!
'Ustle 'im, shake 'im till 'e's sick!
  Wot, 'e *would*, would 'e? Well,
  Then yer've got ter give 'im 'Ell,
An' it's trunch, trunch, truncheon does the trick!

147 *A Luncheon (Thomas Hardy entertains the Prince of Wales)*

LIFT latch, step in, be welcome, Sir,
Albeit to see you I'm unglad
And your face is fraught with a deathly shyness
Bleaching what pink it may have had,
Come in, come in, Your Royal Highness.

Beautiful weather?—Sir, that's true,
Though the farmers are casting rueful looks
At tilth's and pasture's dearth of spryness.—
Yes, Sir, I've written several books.—
A little more chicken, Your Royal Highness?

Lift latch, step out, your car is there,
To bear you hence from this antient vale.
We are both of us aged by our strange brief nighness,
But each of us lives to tell the tale.
Farewell, farewell, Your Royal Highness.

WALTER DE LA MARE
1873–1956

148 *Iron*

IT is the gentle poet's art
In pleasing diction to impart
  Whatever he thinks meet:
And even make the ugly bloom
  In splendour at our feet.
But neither Shelley, Keats nor Byron
Sang songs on Zinc, or odes to Iron:
  Impracticable feat!

When passing, then, I always bow
To him who makes (I know not how)
A living out of nails, pans, pails—
  I bow across the street—
Just bow: and then my courage fails:
  I beat a swift retreat.

For who can help but ponder on
His awful state when, Sunday gone,
  At daybreak bleak and chill,
He turns the shop-key in its lock,
Stares in upon his ghastly stock
  And opens Monday's till?

149

The Bards

MY agèd friend, Miss Wilkinson,
  Whose mother was a Lambe,
Saw Wordsworth once, and Coleridge, too,
  One morning in her p'ram.[1]

Birdlike the bards stooped over her—
  Like fledgling in a nest;
And Wordsworth said, 'Thou harmless babe!'
  And Coleridge was impressed.

The pretty thing gazed up and smiled,
  And softly murmured, 'Coo!'
William was then aged sixty-four
  And Samuel sixty-two.

[1] This was a three-wheeled vehicle
  Of iron and of wood;
It had a leather apron,
  But it hadn't any hood.

G. K. CHESTERTON
1874–1936

150

Triolet

I WISH I were a jelly fish
That cannot fall downstairs:
Of all the things I wish to wish
I wish I were a jelly fish
That hasn't any cares,
And doesn't even have to wish
'I wish I were a jelly fish
That cannot fall downstairs.'

151

Variations on an Air composed on having to appear in a pageant as Old King Cole

OLD King Cole was a merry old soul,
And a merry old soul was he;
He called for his pipe,
He called for his bowl,
And he called for his fiddlers three.

After Lord Tennyson

Cole, that unwearied prince of Colchester,
Growing more gay with age and with long days
Deeper in laughter and desire of life,
As that Virginian climber on our walls
Flames scarlet with the fading of the year;
Called for his wassail and that other weed
Virginian also, from the western woods
Where English Raleigh checked the boast of Spain,
And lighting joy with joy, and piling up
Pleasure as crown for pleasure, bade men bring
Those three, the minstrels whose emblazoned coats

Shone with the oyster-shells of Colchester;
And these three played, and playing grew more fain
Of mirth and music; till the heathen came,
And the King slept beside the northern sea.

After W. B. Yeats

Of an old King in a story
  From the grey sea-folk I have heard,
Whose heart was no more broken
  Than the wings of a bird.

As soon as the moon was silver
  And the thin stars began,
He took his pipe and his tankard,
  Like an old peasant man.

And three tall shadows were with him
  And came at his command;
And played before him for ever
  The fiddles of fairyland.

And he died in the young summer
  Of the world's desire;
Before our hearts were broken
  Like sticks in a fire.

After Robert Browning

Who smoke-snorts toasts o' My Lady Nicotine,
Kicks stuffing out of Pussyfoot, bids his trio
Stick up their Stradivarii (that's the plural)
Or near enough, my fatheads; *nimium*
Vicina Cremonæ (that's a bit too near).
Is there some stockfish fails to understand?
Catch hold o' the notion, bellow and blurt back 'Cole'?
Must I bawl lessons from a horn-book, howl,
Cat-call the cat-gut 'fiddles'? Fiddlesticks!

G. K. CHESTERTON

After Walt Whitman

Me clairvoyant,
Me conscious of you, old camarado,
Needing no telescope, lorgnette, field-glass, opera-glass, myopic pince-nez,
Me piercing two thousand years with eye naked and not ashamed;
The crown cannot hide you from me;
Musty old feudal-heraldic trappings cannot hide you from me,
I perceive that you drink.
(I am drinking with you. I am as drunk as you are.)
I see you are inhaling tobacco, puffing, smoking, spitting
(I do not object to your spitting),
You prophetic of American largeness,
You anticipating the broad masculine manners of these States;
I see in you also there are movements, tremors, tears, desire for the melodious,
I salute your three violinists, endlessly making vibrations,
Rigid, relentless, capable of going on for ever;
They play my accompaniment; but I shall take no notice of any accompaniment;
I myself am a complete orchestra.
So long.

After Swinburne

In the time of old sin without sadness
  And golden with wastage of gold,
Like the gods that grow old in their gladness
  Was the king that was glad, growing old;
And with sound of loud lyres from his palace
  The voice of his oracles spoke,
And the lips that were red from his chalice
  Were splendid with smoke.

When the weed was as flame for a token
  And the wine was as blood for a sign;
And upheld in his hands and unbroken
  The fountains of fire and of wine.
And a song without speech, without singer,
  Stung the soul of a thousand in three
As the flesh of the earth has to sting her,
  The soul of the sea.

152 *A Ballad of Abbreviations*

THE American's a hustler, for he says so,
  And surely the American must know.
He will prove to you with figures why it pays so,
  Beginning with his boyhood long ago.
When the slow-maturing anecdote is ripest,
  He'll dictate it like a Board of Trade Report,
And because he has no time to call a typist,
  He calls her a Stenographer for short.

He is never known to loiter or malinger,
  He rushes, for he knows he has 'a date';
He is always on the spot and full of ginger,
  Which is why he is invariably late.
When he guesses that it's getting even later,
  His vocabulary's vehement and swift,
And he yells for what he calls the Elevator,
  A slang abbreviation for a lift.

Then nothing can be nattier or nicer
  For those who like a light and rapid style,
Than to trifle with a work of Mr. Dreiser
  As it comes along in waggons by the mile.
He had taught us what a swift selective art meant
  By description of his dinners and all that,
And his dwelling, which he says is an Apartment,
  Because he cannot stop to say a flat.

We may whisper of his wild precipitation,
  That its speed is rather longer than a span,
But there really is a definite occasion
  When he does not use the longest word he can.
When he substitutes, I freely make admission,
  One shorter and much easier to spell;
If you ask him what he thinks of Prohibition,
  He may tell you quite succinctly it is Hell.

153 *Antichrist, or the Reunion of Christendom: An Ode*

"A Bill which has shocked the conscience of every Christian community in Europe"—*Mr. F. E. Smith, on the Welsh Disestablishment Bill.*

ARE they clinging to their crosses,
F. E. Smith,
Where the Breton boat-fleet tosses,
Are they, Smith?
Do they, fasting, trembling, bleeding,
Wait the news from this our city?
Groaning 'That's the Second Reading!'
Hissing 'There is still Committee!'
If the voice of Cecil falters,
If McKenna's point has pith
Do they tremble for their altars?
Do they, Smith?

Russian peasants round their pope
Huddled, Smith,
Hear about it all, I hope,
Don't they, Smith?
In the mountain hamlets clothing
Peaks beyond Caucasian pales,
Where Establishment means nothing
And they never heard of Wales,
Do they read it all in Hansard
With a crib to read it with—
'Welsh Tithes: Dr. Clifford Answered.'
Really, Smith?

In the lands where Christians were,
F. E. Smith,
In the little lands laid bare,
Smith, O Smith!
Where the Turkish bands are busy,
And the Tory name is blessed
Since they hailed the Cross of Dizzy
On the banners from the West!
Men don't think it half so hard if
Islam burns their kin and kith,
Since a curate lives in Cardiff
Saved by Smith.

It would greatly, I must own,
Soothe me, Smith!
If you left this theme alone,
Holy Smith!
For your legal cause or civil
You fight well and get your fee;
For your God or dream or devil
You will answer, not to me.
Talk about the pews and steeples
And the Cash that goes therewith!
But the souls of Christian peoples . . .
Chuck it, Smith!

154 *The Rolling English Road*

BEFORE the Roman came to Rye or out to Severn strode,
The rolling English drunkard made the rolling English road.
A reeling road, a rolling road, that rambles round the shire,
And after him the parson ran, the sexton and the squire;
A merry road, a mazy road, and such as we did tread
That night we went to Birmingham by way of Beachy Head.

I knew no harm of Bonaparte and plenty of the Squire,
And for to fight the Frenchmen I did not much desire;
But I did bash their baggonets because they came arrayed
To straighten out the crooked road an English drunkard made,
Where you and I went down the lane with ale-mugs in our hands,
The night we went to Glastonbury by way of Goodwin Sands.

His sins they were forgiven him; or why do flowers run
Behind him; and the hedges all strengthening in the sun?
The wild thing went from left to right and knew not which was which,
But the wild rose was above him when they found him in the ditch.
God pardon us, nor harden us; we did not see so clear
The night we went to Bannockburn by way of Brighton Pier.

My friends, we will not go again or ape an ancient rage,
Or stretch the folly of our youth to be the shame of age,
But walk with clearer eyes and ears this path that wandereth,
And see undrugged in evening light the decent inn of death;
For there is good news yet to hear and fine things to be seen,
Before we go to Paradise by way of Kensal Green.

ROBERT FROST

1874–1963

155

Departmental

AN ant on the tablecloth
Ran into a dormant moth
Of many times his size.
He showed not the least surprise.
His business wasn't with such.
He gave it scarcely a touch,
And was off on his duty run.
Yet if he encountered one
Of the hive's inquiry squad
Whose work is to find out God
And the nature of time and space,
He would put him onto the case.
Ants are a curious race;
One crossing with hurried tread
The body of one of their dead
Isn't given a moment's arrest—
Seems not even impressed.
But he no doubt reports to any
With whom he crosses antennae,
And they no doubt report
To the higher-up at court.
Then word goes forth in Formic:
'Death's come to Jerry McCormic,
Our selfless forager Jerry.
Will the special Janizary
Whose office it is to bury
The dead of the commissary
Go bring him home to his people.
Lay him in state on a sepal.
Wrap him for shroud in a petal.
Embalm him with ichor of nettle.
This is the word of your Queen.'
And presently on the scene
Appears a solemn mortician;
And taking formal position

With feelers calmly atwiddle,
Seizes the dead by the middle,
And heaving him high in air,
Carries him out of there.
No one stands round to stare.
It is nobody else's affair.

It couldn't be called ungentle.
But how thoroughly departmental.

E. C. BENTLEY

1875–1956

156

Clerihews

(i)

THE Art of Biography
Is different from Geography.
Geography is about Maps,
But Biography is about Chaps.

(ii)

WHAT I like about Clive
Is that he is no longer alive.
There is a great deal to be said
For being dead.

(iii)

I AM *not* Mahomet.
—Far from it.
That is the mistake
All of you seem to make.

(iv)

GEORGE the Third
Ought never to have occurred.
One can only wonder
At so grotesque a blunder.

(v)

THERE exists no proof as
To who shot William Rufus,
But shooting him would seem
To have been quite a sound scheme.

(vi)

'No,' said Charles Peace,
'I can't 'ardly blame the perlice.
They 'as their faults, it is true,
But I sees their point of view.'

(vii)

'No, sir,' said General Sherman,
'I did *not* enjoy the sermon;
Nor I didn't git any
Kick outer the Litany.'

WALLACE IRWIN

1875–1959

157

Reminiscence

WHEN many years we'd been apart
I met Sad Jim ashore
And set to talkin' heart to heart
About the days of yore.

'Do you recall them happy days?'
'I don't,' says Jim, 'do you?'
I speaks up hearty and I says,
'Be jiggered if I do!'

'Then why are you recallin' of
The joyful days gone by,
The songs and girls we ust to love?'
'What songs and girls?' says I.

'I guess I have fergot,' says Jim
  And started N N E.
It seems I had the best o' him
  And him the best o' me.

E. V. KNOX

1881–1971

158 *To the God of Love*

COME to me, Eros, if you needs must come
  This year, with milder twinges;
Aim not your arrow at the bullseye plumb,
But let the outer pericardium
  Be where the point impinges.

Garishly beautiful I watch them wane
  Like sunsets in a pink west,
The passions of the past; but O their pain!
You recollect that nice affair with Jane?
  We nearly had an inquest.

I want some mellower romance than these,
  Something that shall not waken
The bosom of the bard from midnight ease,
Nor spoil his appetite for breakfast, please
  (Porridge and eggs and bacon).

Something that shall not steep the soul in gall,
  Nor plant it *in excelsis*,
Nor quite prevent the bondman in its thrall
From biffing off the tee as good a ball
  As anybody else's;

But rather, when the world is dull and gray
  And everything seems horrid,
And books are impotent to charm away
The leaden-footed hours, shall make me say,
  'My hat!' (and strike my forehead)

'I am in love, O circumstance how sweet!
   O ne'er to be forgot knot!'
And praise the damsel's eyebrows, and repeat
Her name out loud, until it's time to eat,
   Or go to bed, or what not.

This is the kind of desultory bolt,
   Eros, I bid you shoot me;
One with no barb to agitate and jolt,
One where the feathers have begun to moult—
   Any old sort will suit me.

P. G. WODEHOUSE

1881–1976

159 *To William (Whom We Have Missed)*

BRIGHT are the days which the Fates hold in store for us,
   BUFFALO BILL, you are with us at last.
Magical name! What a joy it once bore for us!
   How it recalls all the tales of the past,
  Some that we read of in prose or in verse,
  Others, perhaps, which we heard from our nurse.

Tales of the days when to rob and assassinate
   Filled the poor Indian with exquisite glee,
Formed an amusement which ne'er ceased to fascinate,
   Set up his health like a week by the sea.
  Nothing could hinder his playful proclivities,
  Till *you* looked in on the genial festivities.

Then, as a pigeon attempts to fly *from* a hawk,
   Hastily winging its way through the blue,
So did the reveller, dropping his tomahawk,
   Flee at the sight, Colonel CODY, of you.
  Glancing behind with uneasiness palpable,
  Feeling his, too, was a head that was scalpable.

And, at the speed at which lovers, who pant, elope,
    You, with a look both determined and grim,
Covered the ground like an ostrich or antelope,
    Thoroughly bent upon collaring him.
  That was the duty you owed the community,
  Not to allow him to raid with impunity.

Once I considered these tales of your quality
    Nought but a beautiful, wonderful myth,
Scorned to believe that you were, in reality,
    Merely a mortal like BROWN, JONES, and SMITH.
  Briefly, I classed you with ORSON's friend VALENTINE,
  ROBINSON CRUSOE, and heroes of BALLANTYNE.

Now that the years have brought hairs that are silvery,
    Ills that are painful, and views that are fresh,
Only in one thing unchanged, I am still very
    Anxious to look upon you in the flesh.
  Last time I saw you not (owing to gout) at all.
  SQUILLS would not hear of my venturing out at all.

WILLIAM, I'm loth to examine futurity,
    Speak as a prophet regarding your show,
Say if the pageant is doomed to obscurity,
    Or, on the contrary, if it will '*go*',
  Whether 'twill charm or displease, when we view it, us.
  Accurate forecasts are very fortuitous.

Still, when your ochred and plume-covered savages
    Make preparations for raising the hair,
And when your Cowboys are stemming their ravages,
    I, it may please you to know, shall be there.
  One, if no more, of the thousands who pen you in
  Looks on your feats with a pleasure that's genuine.

160

The Gourmet's Love-Song

How strange is Love; I am not one
  Who Cupid's power belittles,
For Cupid 'tis who makes me shun
  My customary victuals.
Oh, Effie, since that painful scene
  That left me broken-hearted,
My appetite, erstwhile so keen,
  Has utterly departed.

My form, my friends observe with pain,
  Is growing daily thinner.
Love only occupies the brain
  That once could think of dinner.
Around me myriad waiters flit,
  With meat and drink to ply men;
Alone, disconsolate, I sit,
  And feed on thoughts of Hymen.

The kindly waiters hear my groan,
  They strive to charm with curry;
They tempt me with a devilled bone—
  I beg them not to worry.
Soup, whitebait, entrées, fricassees,
  They bring me uninvited.
I need them not, for what are these
  To one whose life is blighted?

They show me dishes rich and rare,
  But ah! my pulse no joy stirs.
For savouries I've ceased to care,
  I hate the thought of oysters.
They bring me roast, they bring me boiled,
  But all in vain they woo me;
The waiters softly mutter, 'Foiled!'
  The chef, poor man, looks gloomy.

So, EFFIE, turn that shell-like ear,
  Nor to my sighing close it,
You cannot doubt that I'm sincere—
  This ballad surely shows it.
No longer spurn the suit I press,
  Respect my agitation,
Do change your mind, and answer, 'Yes',
  And save me from starvation.

J. C. SQUIRE

1884–1958

161

A Vision of Truth

As it fell upon a day
I made another garden, yea,
I got me flowers to strew the way
  Like to the summer's rain;
And the chaffinch sings on the orchard bough
'Poor moralist, and what art thou?
But blessings on thy frosty pow,
  And she shall rise again!'

Lord Ullin reached that fatal shore,
A highly respectable Chancellor,
A military casque he wore
  Half-hidden from the eye;
The robin redbreast and the wren,
The Pickwick, the Owl and the Waverley pen,
Heckety-peckety my black hen,
  He took her with a sigh.

The fight is o'er, the battle won,
And furious Frank and fiery Hun,
Stole a pig and away he run
  And drew my snickersnee,
A gulf divides the best and worst
'Ho! bring us wine to quench our thirst!'
We were the first who ever burst
  Under the greenwood tree.

Little Bo-peep fell fast asleep
(She is a shepherdess of sheep),
Bid me to weep and I will weep,
  Thy tooth is not so keen,
Then up and spake Sir Patrick Spens
Who bought a fiddle for eighteenpence
And reverently departed thence,
  His wife could eat no lean.

Epilogue

'Twas roses, roses all the way
  Nor any drop to drink.

RUPERT BROOKE

1887–1915

162

Wagner

CREEPS in half wanton, half asleep,
  One with a fat wide hairless face.
He likes love-music that is cheap;
  Likes women in a crowded place;
    And wants to hear the noise they're making.

His heavy eyelids droop half-over,
  Great pouches swing beneath his eyes.
He listens, thinks himself the lover,
  Heaves from his stomach wheezy sighs;
    He likes to feel his heart's a-breaking.

The music swells. His gross legs quiver.
  His little lips are bright with slime.
The music swells. The women shiver.
  And all the while, in perfect time,
    His pendulous stomach hangs a-shaking.

163 *Sonnet Reversed*

HAND trembling towards hand; the amazing lights
Of heart and eye. They stood on supreme heights.

Ah, the delirious weeks of honeymoon!
  Soon they returned, and, after strange adventures,
Settled at Balham by the end of June.
  Their money was in Can. Pacs. B. Debentures,
And in Antofagastas. Still he went
  Cityward daily; still she did abide
At home. And both were really quite content
  With work and social pleasures. Then they died.
They left three children (besides George, who drank):
  The eldest Jane, who married Mr. Bell,
William, the head-clerk in the County Bank,
  And Henry, a stock-broker, doing well.

T. S. ELIOT

1888–1965

164 *Macavity: The Mystery Cat*

MACAVITY's a Mystery Cat: he's called the Hidden Paw—
For he's the master criminal who can defy the Law.
He's the bafflement of Scotland Yard, the Flying Squad's despair:
For when they reach the scene of crime—*Macavity's not there*!

Macavity, Macavity, there's no one like Macavity,
He's broken every human law, he breaks the law of gravity.
His powers of levitation would make a fakir stare,
And when you reach the scene of crime—*Macavity's not there*!
You may seek him in the basement, you may look up in the air—
But I tell you once and once again, *Macavity's not there*!

Macavity's a ginger cat, he's very tall and thin;
You would know him if you saw him, for his eyes are sunken in.
His brow is deeply lined with thought, his head is highly domed;
His coat is dusty from neglect, his whiskers are uncombed.
He sways his head from side to side, with movements like a snake;
And when you think he's half asleep, he's always wide awake.

Macavity, Macavity, there's no one like Macavity,
For he's a fiend in feline shape, a monster of depravity.
You may meet him in a by-street, you may see him in the square—
But when a crime's discovered, then *Macavity's not there*!

He's outwardly respectable. (They say he cheats at cards.)
And his footprints are not found in any file of Scotland Yard's.
And when the larder's looted, or the jewel-case is rifled,
Or when the milk is missing, or another Peke's been stifled,
Or the greenhouse glass is broken, and the trellis past repair—
Ay, there's the wonder of the thing! *Macavity's not there*!

And when the Foreign Office find a Treaty's gone astray,
Or the Admiralty lose some plans and drawings by the way,
There may be a scrap of paper in the hall or on the stair—
But it's useless to investigate—*Macavity's not there*!
And when the loss has been disclosed, the Secret Service say:
'It *must* have been Macavity!'—but he's a mile away.
You'll be sure to find him resting, or a-licking of his thumbs,
Or engaged in doing complicated long division sums.

Macavity, Macavity, there's no one like Macavity,
There never was a Cat of such deceitfulness and suavity.
He always has an alibi, and one or two to spare:
At whatever time the deed took place—MACAVITY WASN'T THERE!
And they say that all the Cats whose wicked deeds are widely known
(I might mention Mungojerrie, I might mention Griddlebone)
Are nothing more than agents for the Cat who all the time
Just controls their operations: the Napoleon of Crime!

165 *Skimbleshanks: The Railway Cat*

THERE's a whisper down the line at 11.39
When the Night Mail's ready to depart,
Saying 'Skimble where is Skimble has he gone to hunt the thimble?
We must find him or the train can't start.'
All the guards and all the porters and the stationmaster's daughters
They are searching high and low,
Saying 'Skimble where is Skimble for unless he's very nimble
Then the Night Mail just can't go.'
At 11.42 then the signal's nearly due
And the passengers are frantic to a man—
Then Skimble will appear and he'll saunter to the rear:
He's been busy in the luggage van!
  He gives one flash of his glass-green eyes
    And the signal goes 'All Clear!'
  And we're off at last for the northern part
    Of the Northern Hemisphere!

You may say that by and large it is Skimble who's in charge
Of the Sleeping Car Express.
From the driver and the guards to the bagmen playing cards
He will supervise them all, more or less.
Down the corridor he paces and examines all the faces
Of the travellers in the First and in the Third;
He establishes control by a regular patrol
And he'd know at once if anything occurred.
He will watch you without winking and he sees what you are thinking
And it's certain that he doesn't approve
Of hilarity and riot, so the folk are very quiet
When Skimble is about and on the move.
  You can play no pranks with Skimbleshanks!
    He's a Cat that cannot be ignored;
  So nothing goes wrong on the Northern Mail
    When Skimbleshanks is aboard.

Oh it's very pleasant when you have found your little den
With your name written up on the door.
And the berth is very neat with a newly folded sheet
And there's not a speck of dust on the floor.
There is every sort of light—you can make it dark or bright;
There's a handle that you turn to make a breeze.
There's a funny little basin you're supposed to wash your face in
And a crank to shut the window if you sneeze.
Then the guard looks in politely and will ask you very brightly
'Do you like your morning tea weak or strong?'
But Skimble's just behind him and was ready to remind him,
For Skimble won't let anything go wrong.
    And when you creep into your cosy berth
        And pull up the counterpane,
    You ought to reflect that it's very nice
    To know that you won't be bothered by mice—
    You can leave all that to the Railway Cat,
        The Cat of the Railway Train!

In the watches of the night he is always fresh and bright;
Every now and then he has a cup of tea
With perhaps a drop of Scotch while he's keeping on the watch,
Only stopping here and there to catch a flea.
You were fast asleep at Crewe and so you never knew
That he was walking up and down the station;
You were sleeping all the while he was busy at Carlisle,
Where he greets the stationmaster with elation.
But you saw him at Dumfries, where he speaks to the police
If there's anything they ought to know about:
When you get to Gallowgate there you do not have to wait—
For Skimbleshanks will help you to get out!
    He gives you a wave of his long brown tail
        Which says: 'I'll see you again!
    You'll meet without fail on the Midnight Mail
        The cat of the Railway Train.'

166 *Cat Morgan Introduces Himself*

I ONCE was a Pirate what sailed the 'igh seas—
  But now I've retired as a com-mission-aire:
And that's how you find me a-takin' my ease
  And keepin' the door in a Bloomsbury Square.

I'm partial to partridges, likewise to grouse,
  And I favour that Devonshire cream in a bowl;
But I'm allus content with a drink on the 'ouse
  And a bit o' cold fish when I done me patrol.

I ain't got much polish, me manners is gruff,
  But I've got a good coat, and I keep meself smart;
And everyone says, and I guess that's enough:
  'You can't but like Morgan, 'e's got a kind 'art.'

I got knocked about on the Barbary Coast,
  And me voice it ain't no sich melliferous horgan;
But yet I can state, and I'm not one to boast,
  That some of the gals is dead keen on old Morgan.

So if you 'ave business with Faber—or Faber—
  I'll give you this tip, and it's worth a lot more:
You'll save yourself time, and you'll spare yourself
    labour
  If jist you make friends with the Cat at the door.

—Morgan

HUGH KINGSMILL

1889–1949

167

Two Poems
(*after A. E. Housman*)

(i)

WHAT, still alive at twenty-two,
A clean upstanding chap like you?
Sure, if your throat 'tis hard to slit,
Slit your girl's, and swing for it.

Like enough, you won't be glad,
When they come to hang you, lad:
But bacon's not the only thing
That's cured by hanging from a string.

So, when the spilt ink of the night
Spreads o'er the blotting pad of light,
Lads whose job is still to do
Shall whet their knives, and think of you.

(ii)

'TIS Summer Time on Bredon,
  And now the farmers swear;
The cattle rise and listen
  In valleys far and near,
  And blush at what they hear.

But when the mists in autumn
  On Bredon tops are thick,
The happy hymns of farmers
  Go up from fold and rick,
  The cattle then are sick.

168

Limericks

(i)

THERE was an old man of Khartoum
Who kept a tame sheep in his room,
  'To remind me,' he said,
  'Of someone who's dead,
But I never can recollect whom.'

W. R. INGE (1860–1954)

(ii)

THERE once was a man who said, 'Damn!
It is borne in upon me I am
  An engine that moves
  In predestinate grooves:
I'm not even a bus I'm a tram.'

M. E. HARE (1886–1967)

(iii)

THERE was a young lady named Bright
Whose speed was far faster than light;
  She set out one day
  In a relative way,
And returned home the previous night.

ARTHUR BULLER (1874–1944)

(iv)

THERE once was a man who said 'God
Must think it exceedingly odd
  If he finds that this tree
  Continues to be
When there's no one about in the Quad.'

R. A. KNOX (1888–1957)

(v)

DEAR Sir, Your astonishment's odd:
I am always about in the Quad.
  And that's why the tree
  Will continue to be,
Since observed by Yours faithfully, God.

(vi)

THERE was a young woman called Starky
Who had an affair with a darky.
  The result of her sins
  Was quadruplets, not twins:
One black, and one white, and two khaki.

(vii)

THE young things who frequent picture-palaces
Have no time for psycho-analysis,
  And though Dr. Freud
  Is distinctly annoyed
They cling to their long-standing fallacies.

(viii)

THERE'S a notable family named Stein,
There's Gert and there's Ep and there's Ein.
  Gert's prose is all bunk,
  Ep's sculpture just junk,
And nobody understands Ein.

ANONYMOUS

J. R. POPE

?–1941

169

A Word of Encouragement

O WHAT a tangled web we weave
When first we practise to deceive!
But when we've practised quite a while
How vastly we improve our style!

A. P. HERBERT

1890–1971

170

Lines for a Worthy Person who has drifted by accident into a Chelsea revel

IT is a very curious fact
That those who write or paint or act,
    Compose or etch
    Or sculp or sketch,
  Or practise things like pottery,
Have not got consciences like us,
Are frankly not monogamous;
    Their moral tone
    Is all their own,
  Their love-affairs a lottery.
It's hard to say why writing verse
Should terminate in drink or worse,
    Why flutes and harps
    And flats and sharps
  Should lead to indiscretions;
But if you read the Poets' Lives
You'll find the number of their wives
    In fact exceeds
    The normal needs
  Of almost all professions.

As my poor father used to say
In 1863,
Once people start on all this Art
Goodbye, moralitee!
And what my father used to say
Is good enough for me.

Oh, may no little child of mine
Compose or model, draw, design,
And sit at ease
On people's knees,
With other odious habits!
See what eccentric things they wear,
Observe their odd un-English hair—
The women bald,
The men (so-called)
As thickly furred as rabbits!
Not these the kind of people who
Were prominent at Waterloo,
Not this the stock
Which stood the shock
When Kaiser picked his quarrel.
Let Dagoes paint and write and sing,
But Art is not an English thing;
Better be pure
And die obscure
Than famous but immoral!

As my poor father used to say
In 1863,
Once people start on all this Art
Farewell, monogamee!
And what my father used to say,
And what my father used to say,
Is good enough for me.

And shall we let this canker stick
Inside the body politic?
Oh, let us take
Some steps to make
Our messy nation cleaner!
The whole is greater than the part,
We should at once prohibit Art,
Let Music be
A felony
And Verse a misdemeanour;
Let long-haired gentlemen who draw
Be segregated by the law,
And every bard
Do six months' hard
Who lyrically twaddles,
But licences be issued to
A few selected curates, who
Shall fashion odes
In serious modes
On statutory models.

As my poor father used to say
In 1863,
Once people start on all this Art
Farewell, moralitee!
And what my father used to say,
And what my father used to say,
And what my father used to say,
Is good enough for me.

D. B. WYNDHAM LEWIS

(Timothy Shy)

1891–1969

171 *Sapphics*

EXQUISITE torment, dainty Mrs. Hargreaves
Trips down the High Street, slaying hearts a-plenty;
Stricken and doomed are all who meet her eye-shots!
Bar Mr. Hargreaves.

Grocers a-tremble bash their brassy scales down,
Careless of weight and hacking cheese regardless;
Postmen shoot letters in the nearest ashcan,
Dogs dance in circles.

Leaving their meters, gas-inspectors gallop,
Water Board men cease cutting off the water;
Florists are strewing inexpensive posies
In Beauty's pathway.

'O cruel fair!' groan butchers at their chopping,
'Vive la belle Hargreaves!' howls a pallid milkman;
Even the Vicar shades his eyes and mutters:
'*O dea certe.*'

Back to 'Balmoral' trips the goddess lightly;
Night comes at length, and Mr. Hargreaves with it,
Casting his bowler glumly on the sideboard:
'Gimme my dinner.'

J. B. MORTON
(Beachcomber)
1893–

172 *The Dancing Cabman*

ALONE on the lawn
  The cabman dances:
In the dew of dawn
  He kicks and prances.
His bowler is set
  On his bullet-head.
For his boots are wet,
  And his aunt is dead.
There on the lawn,
  As the light advances,
On the tide of the dawn,
  The cabman dances.
Swift and strong
  As a garden roller,
He dances along
  In his little bowler,
Skimming the lawn
  With royal grace,
The dew of the dawn
  On his great red face.
To fairy flutes,
  As the light advances,
In square black boots
  The cabman dances.

COLIN ELLIS

1895–1949

173

The New Vicar of Bray

IN Queen Victoria's early days,
  When Grandpapa was Vicar,
The squire was worldly in his ways,
  And far too fond of liquor.
My grandsire laboured to exhort
  This influential sinner,
As to and fro they passed the port
  On Sunday after dinner.

My father stepped Salvation's road
  To tunes of Tate and Brady's;
His congregation overflowed
  With wealthy maiden ladies.
Yet modern thought he did not shirk—
  He made his contribution
By writing that successful work,
  'The Church and Evolution'.

When I took orders, war and strife
  Filled parsons with misgiving,
For none knew who might lose his life
  Or who might lose his living.
But I was early on the scenes,
  Where some were loth to go, sir!
And there by running Base Canteens
  I won the D.S.O., sir!

You may have read 'The Very Light'—
  A book of verse that I penned—
The proceeds of it, though but slight,
  Eked out my modest stipend.
My grandsire's tactics long had failed,
  And now my father's line did;
So on another tack I sailed
  (You can't be too broad-minded).

The public-house is now the place
  To get to know the men in,
And if the King is in disgrace
  Then I shall shout for Lenin!
And though my feelings they may shock,
  By murder, theft and arson,
The parson still shall keep his flock
  While they will keep the parson!

And this is the law that I'll maintain
  Until my dying day, sir!
That whether King or Mob shall reign,
  I'm for the people that pay, sir!

ROBERT GRAVES

1895–

174 *Epitaph on an Unfortunate Artist*

He found a formula for drawing comic rabbits:
This formula for drawing comic rabbits paid,
So in the end he could not change the tragic habits
This formula for drawing comic rabbits made.

175 *Wm. Brazier*

At the end of Tarriers' Lane, which was the street
We children thought the pleasantest in Town
Because of the old elms growing from the pavement
And the crookedness, when the other streets were straight,
[They were always at the lamp-post round the corner,
Those pugs and papillons and in-betweens,
Nosing and snuffling for the latest news]
Lived Wm. Brazier, with a gilded sign,
'Practical Chimney Sweep'. He had black hands,
Black face, black clothes, black brushes and white teeth;
He jingled round the town in a pony-trap,

And the pony's name was Soot, and Soot was black.
But the brass fittings on the trap, the shafts,
On Soot's black harness, on the black whip-butt,
Twinkled and shone like any guardsman's buttons.
Wasn't that pretty? And when we children jeered:
'Hello, Wm. Brazier! Dirty-face Wm. Brazier!'
He would crack his whip at us and smile and bellow,
'Hello, my dears!' [If he was drunk, but otherwise:
'Scum off, you damned young milliners' bastards, you!']

Let them copy it out on a pink page of their albums,
Carefully leaving out the bracketed lines.
It's an old story—f's for s's—
But good enough for them, the suckers.

176 *The Persian Version*

TRUTH-loving Persians do not dwell upon
The trivial skirmish fought near Marathon.
As for the Greek theatrical tradition
Which represents that summer's expedition
Not as a mere reconnaissance in force
By three brigades of foot and one of horse
(Their left flank covered by some obsolete
Light craft detached from the main Persian fleet)
But as a grandiose, ill-starred attempt
To conquer Greece—they treat it with contempt;
And only incidentally refute
Major Greek claims, by stressing what repute
The Persian monarch and the Persian nation
Won by this salutary demonstration:
Despite a strong defence and adverse weather
All arms combined magnificently together.

177 *'¡Wellcome, to the Caves of Artá!'*

'They are hollowed out in the see coast at the muncipal terminal of Capdepera, at nine kilometer from the town of Artá in the Island of Mallorca, with a suporizing infinity of graceful colums of 21 meter and by downward, which prives the spectator of all animacion and plunges in dumbness. The way going is very picturesque, serpentine between style mountains, til the arrival at the esplanade of the vallee, called 'The Spider'. There are good enlacements of the railroad with autobuses of excursion, many days of the week, today actually Wednesday and Satturday. Since many centuries renown foreing visitors have explored them and wrote their eulogy about, included Nort-American geoglogues.' [*From a Tourist leaflet*] 180

SUCH subtile filigranity and nobless of construccion
  Here fraternise in harmony, that respiracion stops.
While all admit their impotence (though autors most formidable)
  To sing in words the excellence of Nature's underprops,
Yet stalactite and stalagmite together with dumb language
  Make hymnes to God wich celebrate the strength of water drops.

¿You, also, are you capable to make precise in idiom
  Consideracions magic of ilusions very wide?
Alraedy in the Vestibule of these Grand Caves of Artá
  The spirit of the human verb is darked and stupefyed;
So humildy you trespass trough the forest of the colums
  And listen to the grandess explicated by the guide.

From darkness into darkness, but at measure, now descending
  You remark with what esxactitude he designates each bent;
'The Saloon of Thousand Banners', or 'The Tumba of Napoleon',
  'The Grotto of the Rosary', 'The Club', 'The Camping Tent'.
And at 'Cavern of the Organ' there are knocking streange formacions
  Wich give a nois particular pervoking wonderment.

¡Too far do not adventure, sir! For, further as you wander,
  The every of the stalactites will make you stop and stay.
Grand peril amenaces now, your nostrills aprehending
  An odour least delicious of lamentable decay.
It is some poor touristers, in the depth of obscure cristal,
  Wich deceased of thier emocion on a past excursion day.

178

Wigs and Beards

IN the bad old days a bewigged country Squire
Would never pay his debts, unless at cards,
Shot, angled, urged his pack through standing grain,
Horsewhipped his tenantry, snorted at the arts,
Toped himself under the table every night,
Blasphemed God with a cropful of God-damns,
Aired whorehouse French or lame Italian,
Set fashions of pluperfect slovenliness
And claimed seigneurial rights over all women
Who slept, imprudently, under the same roof.

Taxes and wars long ago ploughed them under—
'And serve the bastards right' the Beards agree,
Hurling their empties through the café window
And belching loud as they proceed downstairs.
Latter-day bastards of that famous stock,
That never rode a nag, nor gaffed a trout,
Nor winged a pheasant, nor went soldiering,
But remain true to the same hell-fire code
In all available particulars
And scorn to pay their debts even at cards.
Moreunder (which is to subtract, not add),
Their ancestors called themselves gentlemen
As they, in the same sense, call themselves artists.

L. A. G. STRONG

1896–1958

179

A Memory

WHEN I was as high as that
I saw a poet in his hat.
I think the poet must have smiled
At such a solemn gazing child.

Now wasn't it a funny thing
To get a sight of J. M. Synge,
And notice nothing but his hat?
Yet life is often queer like that.

C. S. LEWIS

1898–1963

Evolutionary Hymn

LEAD us, Evolution, lead us
  Up the future's endless stair;
Chop us, change us, prod us, weed us.
  For stagnation is despair:
Groping, guessing, yet progressing,
  Lead us nobody knows where.

Wrong or justice in the present,
  Joy or sorrow, what are they
While there's always jam to-morrow,
  While we tread the onward way?
Never knowing where we're going,
  We can never go astray.

To whatever variation
  Our posterity may turn
Hairy, squashy, or crustacean,
  Bulbous-eyed or square of stern,
Tusked or toothless, mild or ruthless,
  Towards that unknown god we yearn.

Ask not if it's god or devil,
  Brethren, lest your words imply
Static norms of good and evil
  (As in Plato) throned on high;
Such scholastic, inelastic,
  Abstract yardsticks we deny.

Far too long have sages vainly
  Glossed great Naturc's simplc tcxt;
He who runs can read it plainly,
  'Goodness = what comes next.'
By evolving, Life is solving
  All the questions we perplexed.

On thcn! Valuc means survival-
  Value. If our progeny
Spreads and spawns and licks each rival,
  That will prove its deity
(Far from pleasant, by our present
  Standards, though it well may be).

NOËL COWARD

1899–1973

181 *Mad Dogs and Englishmen*

In tropical climes there are certain times of day,
When all the citizens retire
To tear their clothes off and perspire.
It's one of those rules that the greatest fools obey,
Because the sun is much too sultry
And one must avoid its ultry-violet ray. . . .
The natives grieve when the white men leave their huts,
Because they're ooviously definitely nuts!

Mad dogs and Englishmen
Go out in the midday sun.
The Japanese don't care to,
The Chinese wouldn't dare to,
Hindoos and Argentines sleep firmly from twelve to one,
But Englishmen detest a
Siesta.
In the Philippines there are lovely screens
To protect you from the glare.
In the Malay States there are hats like plates
Which the Britishers won't wear.

At twelve noon
The natives swoon
And no further work is done,
But mad dogs and Englishmen
Go out in the midday sun.

It's such a surprise for the Eastern eyes to see,
That though the English are effete
They're quite impervious to heat.
When the white man rides every native hides in glee,
Because the simple creatures hope he
Will impale his solar topee on a tree. . . .
It seems such a shame when the English claim the earth
That they give rise to such hilarity and mirth.

Mad dogs and Englishmen
Go out in the midday sun.
The toughest Burmese bandit
Can never understand it.
In Rangoon the heat of noon
Is just what the natives shun.
They put their Scotch or rye down
And lie down.
In a jungle town
Where the sun beats down
To the rage of man and beast,
The English garb
Of the English sahib
Merely gets a bit more creased.
In Bangkok
At twelve o'clock
They foam at the mouth and run,
But mad dogs and Englishmen
Go out in the midday sun.

Mad dogs and Englishmen
Go out in the midday sun.
The smallest Malay rabbit
Deplores this foolish habit.
In Hong Kong
They strike a gong
And fire off a noonday gun,
To reprimand each inmate
Who's in late.
In the mangrove swamps
Where the python romps
There is peace from twelve to two.
Even caribous
Lie around and snooze,
For there's nothing else to do.
In Bengal
To move at all
Is seldom, if ever done,
But mad dogs and Englishmen
Go out in the midday sun.

182 *There are Bad Times Just Around the Corner*

THEY'RE out of sorts in Sunderland,
They're terribly cross in Kent,
They're dull in Hull
And the Isle of Mull
Is seething with discontent;
They're nervous in Northumberland
And Devon is down the drain,
They're filled with wrath
In the Firth of Forth
And sullen on Salisbury Plain.
In Dublin they're depressed, lads,
Mainly because they're Celts,
For Drake is going West, lads,
And so is everyone else.
Hurray-hurray-hurray!
Misery's here to stay.

There are bad times just around the corner,
There are dark clouds hurtling through the sky,
And it's no good whining
About a silver lining
For we know from experience that they won't roll by.
With a scowl and frown
We'll keep our peckers down,
And prepare for depression and doom and dread,
We're going to unpack our troubles from our old kitbag
And wait until we drop down dead.

From Portland Bill to Scarborough
They're querulous and subdued
And Shropshire lads
Have behaved like cads
From Berwick-on-Tweed to Bude,
They're mad at Market Harborough
And livid at Leigh-on-Sea,
In Tunbridge Wells
You can hear the yells
Of woe-begone bourgeoisie.
We all get bitched about, lads,
Whoever our vote elects,
We know we're up the spout, lads,
And that's what England expects.
Hurray-hurray-hurray!
Trouble is on the way.

There are bad times just around the corner,
The horizon's gloomy as can be;
There are black birds over
The greyish cliffs of Dover,
And the rats are preparing to leave the BBC.
We're an unhappy breed
And very bored indeed
When reminded of something that Nelson said,
And while the press and the politicians nag nag nag
We'll wait until we drop down dead.

From Colwyn Bay to Kettering
They're sobbing themselves to sleep;
The shrieks and wails
In the Yorkshire dales
Have even depressed the sheep.
In rather vulgar lettering,
A very disgruntled group
Have posted bills
In the Cotswold Hills
To prove that we're in the soup.
While begging Kipling's pardon,
There's one thing we know for sure:
If England is a garden
We ought to have more manure.
Hurray-hurray-hurray!
Suffering and dismay.

There are bad times just around the corner,
And the outlook's absolutely vile;
There are Home Fires smoking
From Windermere to Woking,
And we're *not* going to tighten our belts and smile smile smile.
At the sound of a shot
We'd just as soon as not
Take a hot-water bottle and go to bed:
We're going to *un*tense our muscles till they sag sag sag
And wait until we drop down dead.

There are bad times just around the corner,
We can all look forward to despair,
It's as clear as crystal
From Bridlington to Bristol
That we can't save democracy and we don't much care.
If the Reds and Pinks
Believe that England stinks,
And that world revolution is bound to spread,
We'd better all learn the lyrics of the old 'Red Flag'
And wait until we drop down dead——
A likely story,
Land of Hope and Glory,
Wait until we drop down dead.

ROY CAMPBELL

1902–1957

183

On Some South African Novelists

YOU praise the firm restraint with which they write—
I'm with you there, of course:
They use the snaffle and the curb all right,
But where's the bloody horse?

ANONYMOUS

184

Limericks

(i)

A VICE most obscene and unsavoury
Holds the Bishop of Balham in slavery:
  With maniacal howls
  He rogers young owls
Which he keeps in an underground aviary.

(ii)

A LESBIAN girl of Khartoum
Took a pansy-boy up to her room;
  As they turned out the light
  She said, 'Let's get this right—
Who does what, and with which, and to whom?'

(iii)

AN Argentine gaucho named Bruno
Declared, 'There is one thing I do know:
  A woman is fine
  And a boy is divine,
But a llama is numero uno.'

(iv)

THE breasts of a barmaid of Crale
Were tattooed with the price of brown ale,
While on her behind
For the sake of the blind
Was the same information in Braille.

(v)

WHENEVER a fellow called Rex
Flashed his very small organ of sex,
He always got off,
For the judges would scoff,
'De minimis non curat lex.'

(vi)

WHILE Titian was grinding rose madder
His model was posed on a ladder.
Her position to Titian
Suggested coition
So he dashed up the ladder and had her.

(vii)

THERE was a young Fellow of Wadham
Who asked for a ticket to Sodom.
When they said, 'We prefer
Not to issue them, sir,'
He said, 'Don't call me sir, call me modom.'

(viii)

THERE was a young Fellow of Caius
Who sat with a girl on his knees.
He said to her, 'Miss—
Take more trouble with this,
And pay less attention to these.'

(ix)

THERE was a young Fellow of King's
Who cared not for whores and such things:
His height of desire
Was a boy in the choir
With a bum like a jelly on springs.

185

Clerihew

SPINOZA
Collected curiosa:
Bawdy belles-lettres,
Etc.

186

After Emerson

LIVES of great men all remind us
As we o'er their pages turn,
That we too may leave behind us
Letters that we ought to burn.

STUART HOWARD-JONES

1904–1974

187

Hibernia

MARVELL! I think you'd neither seen nor smelt
In person, the abominable *Celt*.
No sense of shame he'll find, whoever seeks,
Twixt *Kingstown* and the *Macgillicuddy's Reeks*,
As CROMWELL knew, who had these *Apemen* daily
Clocked smartly on the head with a *Shillelagh*.
Observe the Irishman at night in bed:
His bloodshot eyes are popping from his head,
For there he sees, of blackest Midnight born,

A fearful *Banshee* or a *Leprechaun*,
Or other of this pious Country's Demons—
The nat'ral fruit of slight *Delirium Tremens:*
And what's that *Incubus* that looms so big
Upon his chest?—Only—Praise God!—the Pig!

Last night he had put down too much *Potheen*
(A vulgar blend of Methyl and Benzene)
That, at some Wake, he might the better keen.
(Keen—meaning 'brisk'? Nay, here the Language warps:
'Tis singing bawdy Ballads to a Corpse.)

Next day our Irishman goes out of doors
And, pushing through the usual crowd of Whores,
Enters the Church in spirituous Depression
To seek the Priest and make a full Confession:
A useful traffic, for he tells of all
The Tarts he's laid at last night's Funeral,
And thus the Priest can add, perhaps, a name
To his *Compendium* of Easy Game,
While PADRAIC, with his spirit cleansed of sin,
Goes out a better man than he came in.
Then, off to milk his Sow for *Irish Dairies*,
Gets through his penitential *Ave Maries*
And, with a final gabbled *Decalogue*,
Trips and is swallowed in the treach'rous Bog,
And freed from sin, if not precisely shriven,
Goes up to join the BORGIAS in Heaven,
The while his friends, as if their hearts would break,
Lay in the Liquor for another Wake.

Then hear sung, MARVELL, by a wiser Muse,
This land of Popes and Pigs and Bogs and Booze,
For *Lycidas* (and MILTON makes it plain)
Preferred to drown than visit it again.

PHYLLIS McGINLEY

1905–

188

Reflections at Dawn

I WISH I owned a Dior dress
  Made to my order out of satin.
I wish I weighed a little less
  And could read Latin,
Had perfect pitch or matching pearls,
  A better head for street directions,
And seven daughters, all with curls
  And fair complexions.
I wish I'd tan instead of burn.
  But most, on all the stars that glisten,
I wish at parties I could learn
  To sit and listen.

I wish I didn't talk so much at parties.
It isn't that I want *to hear*
My voice assaulting every ear,
Uprising loud and firm and clear
  Above the cocktail clatter.
It's simply, once *a doorbell's rung,*
(*I've been like this since I was young*)
Some madness overtakes my tongue
  And I begin to chatter.

Buffet, ball, banquet, quilting bee,
  Wherever conversation's flowing,
Why must I feel it falls on me
  To keep things going?
Though ladies cleverer than I
  Can loll in silence, soft and idle,
Whatever topic gallops by,
  I seize its bridle,
Hold forth on art, dissect the stage,
  Or babble like a kindergart'ner
Of politics till I enrage
  My dinner partner.

I wish I didn't talk so much at parties.
When hotly boil the arguments,
Ah! would I had the common sense
To sit demurely on a fence
  And let who will be vocal,
Instead of plunging in the fray
With my opinions on display
Till all the gentlemen edge away
  To catch an early local.

Oh! there is many a likely boon
  That fate might flip me from her griddle.
I wish that I could sleep till noon
  And play the fiddle,
Or dance a *tour jeté* so light
  It would not shake a single straw down.
But when I ponder how last night
  I laid the law down,
More than to have the Midas touch
  Or critics' praise, however hearty,
I wish I didn't talk so much,
I wish I didn't talk so much,
I wish I didn't talk so much,
  When I am at a party.

ANTHONY POWELL

1905–

189

Caledonia

PACING with *Bag-Pipe* in a bosky Square,
One morn a *Piper* rent the vernal air,
Dispelling by his savage, baleful Strains
That *Freudian* Pageantry, which Night-time gains.
(He wore a garb deprived of all amenity,
Save for vile jest and *Smoking-Room* obscenity.)
And heark'ning to the *Pibroch's* raucous Note,
Bursting as if from tortur'd porcine Throat,
To Reverie did errant Fancy yield,
Of *Pinkie*, *Solway Moss*, and *Flodden Field*,

And of a Race, whose Thought and Word and Deed
Have made a new INFERNO *North of Tweed*,
Where they can practise in that chilly HELL
Vices that sicken; Virtues that repel.
Did they remain among such unthaw'd Latitudes,
And *Scot* with *Scot* alone exchanged his Platitudes,
Travellers but few would mark their stinted *Bonhomie*,
And smile to see each finicking Œconomy.
Alack! They ever stream through ENGLAND'S door
To batten on the Rich, and grind the Poor,
With furtive Eye and eager, clutching hand
They pass like *Locusts* through the Southern Land:
And line their purses with the yellow Gold,
Which for each Scotchman, London's pavements hold,
And in return, no matter where you find 'em
They brag of Scotland, now left safe behind 'em.

What are this Race whose Pride so rudely burgeons?
Second-rate *Engineers* and obscure *Surgeons*,
Pedant-Philosophers and *Fleet Street* hacks,
With evr'y Quality that *Genius* lacks:
Such Mediocrity was ne'er on view,
Bolster'd by tireless *Scottish Ballyhoo*—
Nay! In two Qualities they stand supreme;
Their *Self-advertisement* and *Self-esteem*.

What of this Land of which they love to talk?
The Sons of SHEM ascend its hills to *stalk*,
And men with nasal *Twang* in *Harris Tweed*
From THE AMERICAS to these succeed.
More rich than CRŒSUS they have cross'd the Main,
To slay the *Grouse* and quaff the dry *Champagne*.
They seek for *Salmon* in *the bonny Brae*,
While *Wall Street* Banks their *bonny* incomes pay;
But Orange-Peel and Cigarette-end ride
On the dank waters of the slimy Clyde
And careful Scots, against a rainy Day
Wade in and bear these Treasures all away.
Better a cottage in the *English* Fens
Than *castellated* Mansions, set in Glens;
The Tents of *Tartars*, or the Huts of *Cossacks*,
Than *Scotch-Baronial* Mansions in the *Trossachs*.

See, where our glorious EMPIRE'S outposts roast
On *Africk's* Sand; or freeze on *Arctick* Coast;
For there a *Scotchman* without doubt you'll view,
Who plagues the *Caffre*; irks the mild *Hindoo*;
And from whose sight chaste *Polynesians* go
As quick as may be; and the *Esquimau*
Grows weary through the endless *Polar* Nights
Of Tales of *Scottish Charm* and *Scotland's Rights*,
Told by some *Scotchman*, who can scarcely stand
In some snow hovel, *North* of *Baffin Land*;
For *Scotchmen*, from *Quebec* to *Kandahar*,
Lay down the Law, while they *prop up* the Bar
In far off lands, beneath th' Imperial Crown
The Bar's *propped up;* THE EMPIRE gets *let down*. . . .
And yet the abject *Picktish* Tribes complain
That ENGLAND'S loss has not proved *Scotland's* gain,
And with each *alcoholick* Breath he draws,
Some *Scotchman* advocates *Disunion's* Cause,
Till *Whisky's* Fumes alone seem to proclaim
The *Pinchbeck* Ornaments of *Scotland's* Name.
And trumpet through the EARTH, The WORLD'S worst News,
The gloomy Glories of the *Scottish Muse*.

Insomnia's Fiend can make no *Scotchman* falter;
He has C-RLY at hand and old Sir W-LT-R:
Books writ by *Scotchmen* and for *Scotchmen* writ,
Devoid of Humour and divorced from Wit,
Well-fitted *Caledonians* to beguile,
In content dull; cacophonous in style.
A Basin! Quick! Ah Me! The Stomach turns!
The Prose of B-RR--, and the Verse of B-RNS!
Can *Scotchmen* still, in face of *Peter Pan*
Deem for a' that, a Man remains a Man?
Rather with B-CH-N'S *Clubmen* let us sport,
Where well-worn *Cliché* apes the *rôle* of Thought;
Or trip through tender *Yarns* by I-N H-Y,
With hoyden Miss and whimsickal R.A.

And there's a Prospect quails the stoutest Heart;
The sombre Gallery of *Scottish* Art:
Soon let us hope the *Cognoscenti* may burn
The pink and pasty Portraiture of R--B-RN
And with the pick and poleaxe rudely hammer on
Those careful *Etchings*, scratched by D.Y. C-M-R-N,
Having destroyed upon the self-same day
Many a *Desert Scene* by J-M-S McB-Y;
While *Highland Criticks* will no more do Battle,
Defending F-RQ-H-RS-N and *Highland Cattle.*

In Musick's Realm this Race (the bitter fact is)
Presume to *teach* an Art they cannot *practise.*
Their *Rhapsodies* and *Rondos*, which abound,
Pollute the Air with *Academick* sound;
But better their morose, pedantic Strain
Than those which to the sterner *North* pertain,
Where ever'y puling *Crofter's* lad of ten
Aspires to be a BARTOK of the Glen.
Belabour'd *Blackamoor* less harshly squeals
Than *Highland Lasses*, dancing *Highland Reels*,
And Ears go numb when *Scottish Neuropaths* play
In Glasgow's Town the *Gaelick Snap* or *Strathspey.*[1]

But *Nationalism's Slogan* sounds at hand
Throughout this sparsely populated Land.
Jacobite Dreams M-CK-NZ--'S Looks engender:
A New *Prince Charlie*? (Or an *Old Pretender*?)
Should ever *Scotchmen Scotchmen* rule again,
For famed M-CD-N-LD they'll not ask in vain;
Fewer such Statesmen ENGLAND might be better for,
Confused of Mind; no less confused of Metaphor.
And cheerless R--TH, unparalleled Enormity
Of *Scottish* Prudery and *Nonconformity*;
Of all his Race scarce one can we desire less,
Teatotal Despot of the ambient *Wireless.*

[1] The lines on music were contributed by the late Constant Lambert.

Let all retrace their steps across the Border!
Yes! Everyone! From L-NKL-T-R to L--D-R!
Back to that *Northern* ATHENS, where men seek,
With *Gaelick* Minds to live a life that's *Greek*.
And Home again, to *Presbyteries* and *Manses*,
Must flock the simpering Tribe of *Scottish Nancies*:
And haunt once more, with Voices high and squeaky,
The foetid lanes and byways of *Auld Reikie*,
And Moor and Mountain, Loch and Glade and Glen,
Take back the *Picti*, fabled Painted Men.

But may *Sweet William* in our Garden grow,
That call to Mind *Culloden* and *Glencoe*,
And when the Night is late and Play is hard,
The *Nine of Diamonds* prove our lucky Card.
Against this Land of *Porridge*, *Scones*, and *Slate*,
Let us rebuild THE WALL, before too late,
To keep for ever from our native Shore
The sottish *Scotchman*, soaked with *Usquebaugh*,
Who fills the mind with phantasies of SADE,
And prayers for some new CUMBERLAND or WADE;
Replace this Race with others, all too few,
The honest *Welchman* or the worthy *Jew*,
The *Ethiop* or Men from far *Cathay*
But let the *Caledonian* stay away.

Such were the Thoughts begotten then and there
By this one *Scotchman*, pacing in the Square.

JOHN BETJEMAN

1906–

190

In Westminster Abbey

LET me take this other glove off
  As the *vox humana* swells,
And the beauteous fields of Eden
  Bask beneath the Abbey bells.
Here, where England's statesmen lie,
Listen to a lady's cry.

Gracious Lord, oh bomb the Germans.
  Spare their women for Thy Sake,
And if that is not too easy
  We will pardon Thy Mistake.
But, gracious Lord, whate'er shall be,
Don't let anyone bomb me.

Keep our Empire undismembered
  Guide our Forces by Thy Hand,
Gallant blacks from far Jamaica,
  Honduras and Togoland;
Protect them Lord in all their fights,
And, even more, protect the whites.

Think of what our Nation stands for,
  Books from Boot's and country lanes,
Free speech, free passes, class distinction,
  Democracy and proper drains.
Lord, put beneath Thy special care
One-eighty-nine Cadogan Square.

Although dear Lord I am a sinner,
  I have done no major crime;
Now I'll come to Evening Service
  Whensoever I have the time.
So, Lord, reserve for me a crown,
And do not let my shares go down.

I will labour for Thy Kingdom,
  Help our lads to win the war,
Send white feathers to the cowards,
  Join the Women's Army Corps,
Then wash the Steps around Thy Throne
In the Eternal Safety Zone.

Now I feel a little better,
  What a treat to hear Thy Word,
Where the bones of leading statesmen
  Have so often been interr'd.
And now, dear Lord, I cannot wait
Because I have a luncheon date.

191

A Subaltern's Love-song

MISS J. HUNTER DUNN, Miss J. Hunter Dunn,
Furnish'd and burnish'd by Aldershot sun,
What strenuous singles we played after tea,
We in the tournament—you against me!

Love-thirty, love-forty, oh! weakness of joy,
The speed of a swallow, the grace of a boy,
With carefullest carelessness, gaily you won,
I am weak from your loveliness, Joan Hunter Dunn.

Miss Joan Hunter Dunn, Miss Joan Hunter Dunn,
How mad I am, sad I am, glad that you won.
The warm-handled racket is back in its press,
But my shock-headed victor, she loves me no less.

Her father's euonymus shines as we walk,
And swing past the summer-house, buried in talk,
And cool the verandah that welcomes us in
To the six-o'clock news and a lime-juice and gin.

The scent of the conifers, sound of the bath,
The view from my bedroom of moss-dappled path,
As I struggle with double-end evening tie,
For we dance at the Golf Club, my victor and I.

On the floor of her bedroom lie blazer and shorts
And the cream-coloured walls are be-trophied with sports,
And westering, questioning settles the sun
On your low-leaded window, Miss Joan Hunter Dunn.

The Hillman is waiting, the light's in the hall,
The pictures of Egypt are bright on the wall,
My sweet, I am standing beside the oak stair
And there on the landing's the light on your hair.

By roads 'not adopted', by woodlanded ways,
She drove to the club in the late summer haze,
Into nine-o'clock Camberley, heavy with bells
And mushroomy, pine-woody, evergreen smells.

Miss Joan Hunter Dunn, Miss Joan Hunter Dunn,
I can hear from the car-park the dance has begun.
Oh! full Surrey twilight! importunate band!
Oh! strongly adorable tennis-girl's hand!

Around us are Rovers and Austins afar,
Above us, the intimate roof of the car,
And here on my right is the girl of my choice,
With the tilt of her nose and the chime of her voice,

And the scent of her wrap, and the words never said,
And the ominous, ominous dancing ahead.
We sat in the car park till twenty to one
And now I'm engaged to Miss Joan Hunter Dunn.

192 *Invasion Exercise on the Poultry Farm*

SOFTLY croons the radiogram, loudly hoot the owls,
Judy gives the door a slam and goes to feed the fowls.
Marty rolls a Craven A around her ruby lips
And runs her yellow fingers down her corduroyded hips,
Shuts her mouth and screws her eyes and puffs her fag alight
And hears some most peculiar cries that echo through the night.
Ting-a-ling the telephone, to-whit to-whoo the owls,
Judy, Judy, Judy girl, and have you fed the fowls?
No answer as the poultry gate is swinging there ajar.
Boom the bombers overhead, between the clouds a star,
And just outside, among the arks, in a shadowy sheltered place
Lie Judy and a paratroop in horrible embrace.
Ting-a-ling the telephone. 'Yes, this is Marty Hayne.'
'Have you seen a paratroop come walking down your lane?
He may be on your premises, he may be somewhere near,
And if he is report the fact to Major Maxton-Weir.'
Marty moves in dread towards the window—standing there
Draws the curtain—sees the guilty movement of the pair.[1]
White with rage and lined with age but strong and sturdy still
Marty now co-ordinates her passions and her will,
She will teach that Judy girl to trifle with the heart
And go and kiss a paratroop like any common tart.

[1] These lines in italic are by Henry Oscar.

She switches up the radiogram and covered by the blare
She goes and gets a riding whip and whirls it in the air,
She fetches down a length of rope and rushes, breathing hard
To let the couple have it for embracing in the yard.
Crack! the pair are paralysed. Click! they cannot stir.
Zip! she's trussed the paratroop. There's no embracing *her*.
'Hullo, hullo, hullo, hullo . . . Major Maxton-Weir?
I've trussed your missing paratroop. He's waiting for you here.'

193 *Pot Pourri from a Surrey Garden*

MILES of pram in the wind and Pam in the gorse track,
  Coco-nut smell of the broom, and a packet of Weights
Press'd in the sand. The thud of a hoof on a horse-track—
    A horse-riding horse for a horse-track—
    Conifer county of Surrey approached
  Through remarkable wrought-iron gates.

Over your boundary now, I wash my face in a bird-bath,
  Then which path shall I take? that over there by the pram?
Down by the pond! or—yes, I will take the slippery third path,
    Trodden away with gym shoes,
    Beautiful fir-dry alley that leads
  To the bountiful body of Pam.

Pam, I adore you, Pam, you great big mountainous sports girl,
  Whizzing them over the net, full of the strength of five:
That old Malvernian brother, you zephyr and khaki shorts girl,
    Although he's playing for Woking,
    Can't stand up
  To your wonderful backhand drive.

See the strength of her arm, as firm and hairy as Hendren's;
  See the size of her thighs, the pout of her lips as, cross,
And full of a pent-up strength, she swipes at the rhododendrons,
    Lucky the rhododendrons,
    And flings her arrogant love-lock
  Back with a petulant toss.

Over the redolent pinewoods, in at the bathroom casement,
  One fine Saturday, Windlesham bells shall call:
Up the Butterfield aisle rich with Gothic enlacement,
    Licensed now for embracement,
    Pam and I, as the organ
  Thunders over you all.

194

How to Get On in Society

Originally set as a competition in 'Time and Tide'

PHONE for the fish-knives, Norman,
  As Cook is a little unnerved;
You kiddies have crumpled the serviettes
  And I must have things daintily served.

Are the requisites all in the toilet?
  The frills round the cutlets can wait
Till the girl has replenished the cruets
  And switched on the logs in the grate.

It's ever so close in the lounge, dear,
  But the vestibule's comfy for tea
And Howard is out riding on horseback
  So do come and take some with me.

Now here is a fork for your pastries
  And do use the couch for your feet;
I know what I wanted to ask you—
  Is trifle sufficient for sweet?

Milk and then just as it comes dear?
  I'm afraid the preserve's full of stones;
Beg pardon, I'm soiling the doileys
  With afternoon tea-cakes and scones.

195 *Longfellow's Visit to Venice*

(To be read in a quiet New England accent)

NEAR the celebrated Lido where the breeze is fresh and free
Stands the ancient port of Venice called the City of the Sea.

All its streets are made of water, all its homes are brick and stone,
Yet it has a picturesqueness which is justly all its own.

Here for centuries have artists come to see the vistas quaint,
Here Bellini set his easel, here he taught his School to paint.

Here the youthful Giorgione gazed upon the domes and towers,
And interpreted his era in a way which pleases ours.

A later artist, Tintoretto, also did his paintings here,
Massive works which generations have continued to revere.

Still to-day come modern artists to portray the buildings fair
And their pictures may be purchased on San Marco's famous Square.

When the bell notes from the belfries and the campaniles chime
Still to-day we find Venetians elegantly killing time

In their gilded old palazzos, while the music in our ears
Is the distant band at Florians mixed with songs of gondoliers.

Thus the New World meets the Old World and the sentiments expressed
Are melodiously mingled in my warm New England breast.

196 *Executive*

I AM a young executive. No cuffs than mine are cleaner;
I have a Slimline brief-case and I use the firm's Cortina.
In every roadside hostelry from here to Burgess Hill
The *maîtres d'hôtel* all know me well and let me sign the bill.

You ask me what it is I do. Well actually, you know,
I'm partly a liaison man and partly P.R.O.
Essentially I integrate the current export drive
And basically I'm viable from ten o'clock till five.

For vital off-the-record work—that's talking transport-wise—
I've a scarlet Aston-Martin—and does she go? She flies!
Pedestrians and dogs and cats—we mark them down for slaughter.
I also own a speed-boat which has never touched the water.

She's built of fibre-glass, of course. I call her 'Mandy Jane'
After a bird I used to know—No soda, please, just plain—
And how did I acquire her? Well to tell you about that
And to put you in the picture I must wear my other hat.

I do some mild developing. The sort of place I need
Is a quiet country market town that's rather run to seed.
A luncheon and a drink or two, a little *savoir faire*—
I fix the Planning Officer, the Town Clerk and the Mayor.

And if some preservationist attempts to interfere
A 'dangerous structure' notice from the Borough Engineer
Will settle any buildings that are standing in our way—
The modern style, sir, with respect, has really come to stay.

LOUIS MACNEICE

1907–1963

197 *Bagpipe Music*

IT'S no go the merrygoround, it's no go the rickshaw,
All we want is a limousine and a ticket for the peepshow.
Their knickers are made of crêpe-de-chine, their shoes are made of
  python,
Their halls are lined with tiger rugs and their walls with heads of bison.

John MacDonald found a corpse, put it under the sofa,
Waited till it came to life and hit it with a poker,
Sold its eyes for souvenirs, sold its blood for whisky,
Kept its bones for dumb-bells to use when he was fifty.

It's no go the Yogi-Man, it's no go Blavatsky,
All we want is a bank balance and a bit of skirt in a taxi.

Annie MacDougall went to milk, caught her foot in the heather,
Woke to hear a dance record playing of Old Vienna.
It's no go your maidenheads, it's no go your culture,
All we want is a Dunlop tyre and the devil mend the puncture.

The Laird o' Phelps spent Hogmanay declaring he was sober,
Counted his feet to prove the fact and found he had one foot over.
Mrs. Carmichael had her fifth, looked at the job with repulsion,
Said to the midwife 'Take it away; I'm through with over-production'.

It's no go the gossip column, it's no go the ceilidh,
All we want is a mother's help and a sugar-stick for the baby.

Willie Murray cut his thumb, couldn't count the damage,
Took the hide of an Ayrshire cow and used it for a bandage.
His brother caught three hundred cran when the seas were lavish,
Threw the bleeders back in the sea and went upon the parish.

It's no go the Herring Board, it's no go the Bible,
All we want is a packet of fags when our hands are idle.

It's no go the picture palace, it's no go the stadium,
It's no go the country cot with a pot of pink geraniums,
It's no go the Government grants, it's no go the elections,
Sit on your arse for fifty years and hang your hat on a pension.

It's no go my honey love, it's no go my poppet;
Work your hands from day to day, the winds will blow the profit.
The glass is falling hour by hour, the glass will fall for ever,
But if you break the bloody glass you won't hold up the weather.

198

FROM *Autumn Journal*

SHELLEY and jazz and lieder and love and hymn-tunes
  And day returns too soon;
We'll get drunk among the roses
  In the valley of the moon.
Give me an aphrodisiac, give me lotus,
  Give me the same again;
Make all the erotic poets of Rome and Ionia
  And Florence and Provence and Spain
Pay a tithe of their sugar to my potion
  And ferment my days
With the twang of Hawaii and the boom of the Congo,
  Let the old Muse loosen her stays
Or give me a new Muse with stockings and suspenders
  And a smile like a cat,
With false eyelashes and finger-nails of carmine
  And dressed by Schiaparelli, with a pill-box hat.
Let the aces run riot round Brooklands,
  Let the tape-machines go drunk,
Turn on the purple spotlight, pull out the Vox Humana,
  Dig up somebody's body in a cloakroom trunk.
Give us sensations and then again sensations—
  Strip-tease, fireworks, all-in wrestling, gin;
Spend your capital, open your house and pawn your padlocks,
  Let the critical sense go out and the Roaring Boys come in.
Give me a houri but houris are too easy,
  Give me a nun;
We'll rape the angels off the golden reredos
  Before we're done.

W. H. AUDEN

1907–1973

199

Uncle Henry

WHEN the Flyin' Scot
fills for shootin', I go southward,
wisin' after coffee, leavin'
  Lady Starkie.

Weady for some fun,
visit yearly Wome, Damascus,
in Mowocco look for fwesh a-
-musin' places.

Where I'll find a fwend,
don't you know, a charmin' cweature,
like a Gweek God and devoted:
how delicious!

All they have they bwing,
Abdul, Nino, Manfwed, Kosta:
here's to women for they bear such
lovely kiddies!

200 *The Unknown Citizen*

(To JS/07/M/378
This Marble Monument
Is Erected by the State)

He was found by the Bureau of Statistics to be
One against whom there was no official complaint,
And all the reports on his conduct agree
That, in the modern sense of an old-fashioned word, he was a saint,
For in everything he did he served the Greater Community.
Except for the War till the day he retired
He worked in a factory and never got fired,
But satisfied his employers, Fudge Motors Inc.
Yet he wasn't a scab or odd in his views,
For his Union reports that he paid his dues,
(Our report on his Union shows it was sound)
And our Social Psychology workers found
That he was popular with his mates and liked a drink.
The Press are convinced that he bought a paper every day
And that his reactions to advertisements were normal in every way.
Policies taken out in his name prove that he was fully insured,
And his Health-card shows he was once in hospital but left it cured.
Both Producers Research and High-Grade Living declare
He was fully sensible to the advantages of the Instalment Plan
And had everything necessary to the Modern Man,

A phonograph, a radio, a car and a frigidaire.
Our researchers into Public Opinion are content
That he held the proper opinions for the time of year;
When there was peace, he was for peace; when there was war, he went.
He was married and added five children to the population,
Which our Eugenist says was the right number for a parent of his generation,
And our teachers report that he never interfered with their education.
Was he free? Was he happy? The question is absurd:
Had anything been wrong, we should certainly have heard.

201 FROM *Letter to Lord Byron*

THE thought of writing came to me today
(I like to give these facts of time and space);
The bus was in the desert on its way
From Möthrudalur to some other place:
The tears were streaming down my burning face;
I'd caught a heavy cold in Akureyri,
And lunch was late and life looked very dreary.

Professor Housman was I think the first
To say in print how very stimulating
The little ills by which mankind is cursed,
The colds, the aches, the pains are to creating;
Indeed one hardly goes too far in stating
That many a flawless lyric may be due
Not to a lover's broken heart, but 'flu.

But still a proper explanation's lacking;
Why write to you? I see I must begin
Right at the start when I was at my packing.
The extra pair of socks, the airtight tin
Of China tea, the anti-fly were in;
I asked myself what sort of books I'd read
In Iceland, if I ever felt the need.

I can't read Jefferies on the Wiltshire Downs,
  Nor browse on limericks in a smoking-room;
Who would try Trollope in cathedral towns,
  Or Marie Stopes inside his mother's womb?
  Perhaps you feel the same beyond the tomb.
Do the celestial highbrows only care
For works on Clydeside, Fascists, or Mayfair?

In certain quarters I had heard a rumour
  (For all I know the rumour's only silly)
That Icelanders have little sense of humour.
  I knew the country was extremely hilly,
  The climate unreliable and chilly;
So looking round for something light and easy
I pounced on you as warm and *civilisé*.

There is one other author in my pack:
  For some time I debated which to write to.
Which would least likely send my letter back?
  But I decided that I'd give a fright to
  Jane Austen if I wrote when I'd no right to,
And share in her contempt the dreadful fates
Of Crawford, Musgrave, and of Mr. Yates.

Then she's a novelist. I don't know whether
  You will agree, but novel writing is
A higher art than poetry altogether
  In my opinion, and success implies
  Both finer character and faculties.
Perhaps that's why real novels are as rare
As winter thunder or a polar bear.

The average poet by comparison
  Is unobservant, immature, and lazy.
You must admit, when all is said and done,
  His sense of other people's very hazy,
  His moral judgments are too often crazy,
A slick and easy generalization
Appeals too well to his imagination.

I must remember, though, that you were dead
  Before the four great Russians lived, who brought
The art of novel writing to a head;
  The Book Society had not been bought
  But now the art for which Jane Austen fought,
Under the right persuasion bravely warms
And is the most prodigious of the forms.

She was not an unshockable blue-stocking;
  If shades remain the characters they were,
No doubt she still considers you as shocking.
  But tell Jane Austen, that is, if you dare,
  How much her novels are beloved down here.
She wrote them for posterity, she said;
'Twas rash, but by posterity she's read.

You could not shock her more than she shocks me;
  Beside her Joyce seems innocent as grass.
It makes me most uncomfortable to see
  An English spinster of the middle-class
  Describe the amorous effects of 'brass',
Reveal so frankly and with such sobriety
The economic basis of society.

*

Ottava Rima would, I know, be proper,
  The proper instrument on which to pay,
My compliments, but I should come a cropper;
  Rhyme-royal's difficult enough to play.
  But if no classics as in Chaucer's day,
At least my modern pieces shall be cheery
Like English bishops on the Quantum Theory.

Light verse, poor girl, is under a sad weather;
  Except by Milne and persons of that kind
She's treated as *démodé* altogether.
  It's strange and very unjust to my mind
  Her brief appearances should be confined,
Apart from Belloc's *Cautionary Tales*,
To the more bourgeois periodicals.

'The fascination of what's difficult,'
  The wish to do what one's not done before,
Is, I hope, proper to *Quicunque Vult*,
  The proper card to show at Heaven's door.
  Gerettet not *Gerichtet* be the Law,
Et cetera, et cetera. O curse,
That is the flattest line in English verse.

*

I like your muse because she's gay and witty,
  Because she's neither prostitute nor frump,
The daughter of a European city,
  And country houses long before the slump;
  I like her voice that does not make me jump:
And you I find sympatisch, a good townee,
Neither a preacher, ninny, bore, nor Brownie.

A poet, swimmer, peer, and man of action,
  —It beats Roy Campbell's record by a mile—
You offer every possible attraction.
  By looking into your poetic style
  And love-life on the chance that both were vile,
Several have earned a decent livelihood,
Whose lives were uncreative but were good.

You've had your packet from the critics, though:
  They grant you warmth of heart, but at your head
Their moral and aesthetic brickbats throw.
  A 'vulgar genius' so George Eliot said,
  Which doesn't matter as George Eliot's dead,
But T. S. Eliot, I am sad to find,
Damns you with: 'an uninteresting mind'.

A statement which I must say I'm ashamed at;
  A poet must be judged by his intention,
And serious thought you never said you aimed at.
  I think a serious critic ought to mention
  That one verse style was really your invention,
A style whose meaning does not need a spanner,
You are the master of the airy manner.

By all means let us touch our humble caps to
  La poésie pure, the epic narrative;
But comedy shall get its round of claps, too.
  According to his powers, each may give;
  Only on varied diet can we live.
The pious fable and the dirty story
Share in the total literary glory.

202

Under Which Lyre
A Reactionary Tract for the Times

(*Phi Beta Kappa Poem, Harvard, 1946*)

ARES at last has quit the field,
The bloodstains on the bushes yield
  To seeping showers,
And in their convalescent state
The fractured towns associate
  With summer flowers.

Encamped upon the college plain
Raw veterans already train
  As freshman forces;
Instructors with sarcastic tongue
Shepherd the battle-weary young
  Through basic courses.

Among bewildering appliances
For mastering the arts and sciences
  They stroll or run,
And nerves that steeled themselves to slaughter
Are shot to pieces by the shorter
  Poems of Donne.

Professors back from secret missions
Resume their proper eruditions,
  Though some regret it;
They liked their dictaphones a lot,
They met some big wheels, and do not
  Let you forget it.

But Zeus' inscrutable decree
Permits the will-to-disagree
To be pandemic,
Ordains that vaudeville shall preach
And every commencement speech
Be a polemic.

Let Ares doze, that other war
Is instantly declared once more
'Twixt those who follow
Precocious Hermes all the way
And those who without qualms obey
Pompous Apollo.

Brutal like all Olympic games,
Though fought with smiles and Christian names
And less dramatic,
This dialectic strife between
The civil gods is just as mean,
And more fanatic.

What high immortals do in mirth
Is life and death on Middle Earth;
Their a-historic
Antipathy forever gripes
All ages and somatic types,
The sophomoric

Who face the future's darkest hints
With giggles or with prairie squints
As stout as Cortez,
And those who like myself turn pale
As we approach with ragged sail
The fattening forties.

The sons of Hermes love to play,
And only do their best when they
Are told they oughtn't;
Apollo's children never shrink
From boring jobs but have to think
Their work important.

Related by antithesis,
A compromise between us is
  Impossible;
Respect perhaps but friendship never:
Falstaff the fool confronts forever
  The prig Prince Hal.

If he would leave the self alone,
Apollo's welcome to the throne,
  Fasces and falcons;
He loves to rule, has always done it;
The earth would soon, did Hermes run it,
  Be like the Balkans.

But jealous of our god of dreams,
His common-sense in secret schemes
  To rule the heart;
Unable to invent the lyre,
Creates with simulated fire
  Official art.

And when he occupies a college,
Truth is replaced by Useful Knowledge;
  He pays particular
Attention to Commercial Thought,
Public Relations, Hygiene, Sport,
  In his curricula.

Athletic, extrovert and crude,
For him, to work in solitude
  Is the offence,
The goal a populous Nirvana:
His shield bears this device: *Mens sana*
  Qui mal y pense.

Today his arms, we must confess,
From Right to Left have met success,
  His banners wave
From Yale to Princeton, and the news
From Broadway to the Book Reviews
  Is very grave.

His radio Homers all day long
In over-Whitmanated song
  That does not scan,
With adjectives laid end to end,
Extol the doughnut and commend
  The Common Man.

His, too, each homely lyric thing
On sport or spousal love or spring
  Or dogs or dusters,
Invented by some court-house bard
For recitation by the yard
  In filibusters.

To him ascend the prize orations
And sets of fugal variations
  On some folk-ballad,
While dietitians sacrifice
A glass of prune-juice or a nice
  Marsh-mallow salad.

Charged with his compound of sensational
Sex plus some undenominational
  Religious matter,
Enormous novels by co-eds
Rain down on our defenceless heads
  Till our teeth chatter.

In fake Hermetic uniforms
Behind our battle-line, in swarms
  That keep alighting,
His existentialists declare
That they are in complete despair,
  Yet go on writing.

No matter; He shall be defied;
White Aphrodite is on our side:
  What though his threat
To organize us grow more critical?
Zeus willing, we, the unpolitical,
  Shall beat him yet.

Lone scholars, sniping from the walls
Of learned periodicals,
  Our facts defend,
Our intellectual marines,
Landing in little magazines
  Capture a trend.

By night our student Underground
At cocktail parties whisper round
  From ear to ear;
Fat figures in the public eye
Collapse next morning, ambushed by
  Some witty sneer.

In our morale must lie our strength:
So, that we may behold at length
  Routed Apollo's
Battalions melt away like fog,
Keep well the Hermetic Decalogue,
  Which runs as follows:—

Thou shalt not do as the dean pleases,
Thou shalt not write thy doctor's thesis
  On education,
Thou shalt not worship projects nor
Shalt thou or thine bow down before
  Administration.

Thou shalt not answer questionnaires
Or quizzes upon World-Affairs,
  Nor with compliance
Take any test. Thou shalt not sit
With statisticians nor commit
  A social science.

Thou shalt not be on friendly terms
With guys in advertising firms,
  Nor speak with such
As read the Bible for its prose,
Nor, above all, make love to those
  Who wash too much.

Thou shalt not live within thy means
Nor on plain water and raw greens.
  If thou must choose
Between the chances, choose the odd;
Read *The New Yorker*, trust in God;
  And take short views.

203

On the Circuit

AMONG Pelagian travellers,
Lost on their lewd conceited way
To Massachusetts, Michigan,
Miami or L.A.,

An airborne instrument I sit,
Predestined nightly to fulfil
Columbia-Giesen-Management's
Unfathomable will,

By whose election justified,
I bring my gospel of the Muse
To fundamentalists, to nuns,
To Gentiles and to Jews,

And daily, seven days a week,
Before a local sense has jelled,
From talking-site to talking-site
Am jet-or-prop-propelled.

Though warm my welcome everywhere
I shift so frequently, so fast,
I cannot now say where I was
The evening before last,

Unless some singular event
Should intervene to save the place,
A truly asinine remark,
A soul-bewitching face,

Or blessed encounter, full of joy,
Unscheduled on the Giesen Plan,
With, here, an addict of Tolkien,
There, a Charles Williams fan.

Since Merit but a dunghill is,
I mount the rostrum unafraid:
Indeed, 'twere damnable to ask
If I am overpaid.

Spirit is willing to repeat
Without a qualm the same old talk,
But Flesh is homesick for our snug
Apartment in New York.

A sulky fifty-six, he finds
A change of mealtime utter hell,
Grown far too crotchety to like
A luxury hotel.

The Bible is a goodly book
I always can peruse with zest,
But really cannot say the same
For Hilton's *Be My Guest*,

Nor bear with equanimity
The radio in students' cars,
Musak at breakfast, or—dear God!—
Girl-organists in bars.

Then, worst of all the anxious thought,
Each time my plane begins to sink
And the No Smoking sign comes on:
What will there be to drink?

Is this a milieu where I must
How grahamgreeneish! How infra dig!
Snatch from the bottle in my bag
An analeptic swig?

Another morning comes: I see,
Dwindling below me on the plane,
The roofs of one more audience
I shall not see again.

God bless the lot of them, although
I don't remember which was which:
God bless the U.S.A., so large,
So friendly, and so rich.

204

Doggerel by a Senior Citizen

for Robert Lederer

OUR earth in 1969
Is not the planet I call mine,
The world, I mean, that gives me strength
To hold off chaos at arm's length.

My Eden landscapes and their climes
Are constructs from Edwardian times,
When bath-rooms took up lots of space,
And, before eating, one said Grace.

The automobile, the aeroplane,
Are useful gadgets, but profane:
The enginry of which I dream
Is moved by water or by steam.

Reason requires that I approve
The light-bulb which I cannot love:
To me more reverence-commanding
A fish-tail burner on the landing.

My family ghosts I fought and routed,
Their values, though, I never doubted:
I thought their Protestant Work-Ethic
Both practical and sympathetic.

When couples played or sang duets,
It was immoral to have debts:
I shall continue till I die
To pay in cash for what I buy.

The Book of Common Prayer we knew
Was that of 1662:
Though with-it sermons may be well,
Liturgical reforms are hell.

Sex was, of course—it always is—
The most enticing of mysteries,
But news-stands did not yet supply
Manichaean pornography.

Then Speech was mannerly, an Art,
Like learning not to belch or fart:
I cannot settle which is worse,
The Anti-Novel or Free Verse.

Nor are those Ph.D's my kith,
Who dig the symbol and the myth:
I count myself a man of letters
Who writes, or hopes to, for his betters.

Dare any call Permissiveness
An educational success?
Saner those class-rooms which I sat in,
Compelled to study Greek and Latin.

Though I suspect the term is crap,
If there *is* a Generation Gap,
Who is to blame? Those, old or young,
Who will not learn their Mother-Tongue.

But Love, at least, is not a state
Either *en vogue* or out-of-date,
And I've true friends, I will allow,
To talk and eat with here and now.

Me alienated? Bosh! It's just
As a sworn citizen who must
Skirmish with it that I feel
Most at home with what is Real.

WYNFORD VAUGHAN-THOMAS

1908–

205

Hiraeth in N.W.3

THE sight of the English is getting me down.
Fly westward, my heart, from this festering town
On the Wings of a Dove—and a First Class Return—
To the front room of 'Catref' at Ynys-y-Wern.

Swift through the dark flies the 5.49,
Past Slough and past Didcot and derelict mine,
Past pubs and Lucanias and adverts for ales
Till the back-sides of chapels cry 'Welcome to Wales'.

The lights of the chip-shop shine bright in the dark,
The couples lie laced in the asphalted park;
In the vestry of Carmel (conductor, Seth Hughes)
The iron-lunged Gleemen are raping the Muse.

They're 'Comrades', they're 'Martyrs', they're 'Crossing
  the Plain',
They're roaring of Love in a three-part refrain,
But what hymns from Novello's, at threepence a part,
Can mirror the music I feel in my heart?

Glorious welcome that's waiting for me,
Hymns on the harmonium, Welsh-cakes for tea,
A lecture on Marx: his importance today,
All the raptures of love from a Bangor B.A.!

206

To His Not-So-Coy Mistress

TIME's Wingèd Chariot (poets say)
Warns us to love while yet we may;
Must I not hurry all the more
Who find it parked outside my door?
For those who sipped Love in their prime
Must gulp it down at Closing Time.

207

Farewell to New Zealand

SUPER-SUBURBIA of the Southern Seas,
Nature's—and Reason's—true Antipodes,
Hail, dauntless pioneers, intrepid souls,
Who cleared the Bush—to make a lawn for bowls,
And smashed the noble Maori to ensure
The second-rate were socially secure!
Saved by the Wowsers from the Devil's Tricks,
Your shops, your pubs, your minds all close at six.
Your battle-cry's a deep, contented snore,
You voted Labour, then you worked no more.
The Wharfies' Heaven, the gourmet's Purgat'ry:
Ice-cream on mutton, swilled around in tea!

A Maori fisherman, the legends say,
Dredged up New Zealand in a single day.
I've seen the catch, and here's my parting crack—
It's under-sized; for God's sake throw it back!

OSBERT LANCASTER

1908–

208

FROM *Afternoons with Baedeker*

(i)

Eireann

THE distant Seychelles are not so remote
Nor Ctesiphon as ultimately dead
As this damp square round which tired echoes float
Of something brilliant that George Moore once said:
Where, still, in pitch-pine snugs pale poets quote
Verses rejected by the Bodley Head.
For in this drained aquarium no breeze
Deposits pollen from more fertile shores
Or kills the smell of long unopened drawers
That clings for ever to these dripping trees.
Where Bloom once wandered, gross and ill-at-ease,
Twice-pensioned heroes of forgotten wars
With misplaced confidence demand applause
Shouting stale slogans on the Liffey quays.

(ii)

English

IN 1910 a royal princess
Contracted measles here;
Last spring a pregnant stewardess
Was found beneath the pier;
Her throat, according to the Press,
Was slit from ear to ear.

In all the years that passed between
These two distressing dates
Our only tragedy has been
The raising of the rates,
Though once a flying-bomb was seen
Far out across the straits.

Heard on this coast, the music of the spheres
Would sound like something from *The Gondoliers*.

(iii)
French

I SHALL not linger in that draughty square
Attracted by the art-nouveau hotel
Nor ring in vain the concierge's bell
And then, engulfed by a profound despair
That finds its echo in the passing trains,
Sit drinking in the café, wondering why,
Maddened by love, a butcher at Versailles
On Tuesday evening made to jump his brains.
Nor shall I visit the Flamboyant church,
Three stars in Michelin, yet by some strange fluke
Left unrestored by Viollet-le-Duc,
To carry out some long-desired research.
Too well I know the power to get one down
Exerted by this grey and shuttered town.

(iv)
Manhattan

HERE those of us who really understand
Feel that the past is very close at hand.
In that old brownstone mansion 'cross the way,
Copied from one that she had seen by chance
When on her honeymoon in Paris, France,
Mrs. Van Dryssel gave her great soirées;
And in the chic apartment house next door
J. Rittenhaus the Second lived—and jumped,
The morning after General Motors slumped,
Down from a love-nest on the thirtieth floor.
Tread softly then, for on this holy ground
You'd hear the 'twenties cry from every stone
And Bye-Bye Blackbird on the saxophone
If only History were wired for sound.

PETER DE VRIES

1910–

209

Bacchanal

'COME live with me and be my love,'
He said, in substance. 'There's no vine
We will not pluck the clusters of,
Or grape we will not turn to wine.'

It's autumn of their second year.
Now he, in seasonal pursuit,
With rich and modulated cheer,
Brings home the festive purple fruit;

And she, by passion once demented
—That woman out of Botticelli—
She brews and bottles, unfermented,
The stupid and abiding jelly.

210

Christmas Family Reunion

SINCE last the tutelary hearth
Has seen the bursting pod of kin,
I've thought how good the family mould,
How solid and how genuine.

Now once again the aunts are here,
The uncles, sisters, brothers,
With candy in the children's hair,
The grownups in each other's.

There's talk of saving room for pie;
Grandma discusses her neuralgia.
I long for time to pass, so I
Can think of all this with nostalgia.

STANLEY J. SHARPLESS

1910–

211 *'Go to the ant'*

'GO to the ant, thou sluggard;
Consider her ways, and be wise.'
Well, I've been to the ant, and I'm buggered
If I think it's one up on us guys;
All that rushing about is damn silly,
(And uneconomic. I bet),
I'd rather consider the lily,
It's got Solomon beat—and no sweat.

PAUL DEHN

1912–1976

212 *A Game of Consequences*

COFFEE-CUPS cool on the Vicar's harmonium,
  Clever guests giggle and duffers despond.
Softly, like the patter of mouse-feet, the whisper
  Of Eversharp Pencil on Basildon Bond.

Separate hands scribble separate phrases—
  Innocent, each, as the new-driven snow.
What will they spell, when the paper's unfolded?
  Lucifer, only, and Belial know.

'Ready, Miss Montague? Come, Mr. Jellaby!'
  (Peek at your papers and finger your chins)
'Shy, Mr. Pomfret? You'd rather the Vicar . . . ?
  Oh, good for the Vicar!' The Vicar begins:

'FAT MR. POMFRET met FROWSTY MISS MONTAGUE
  Under the BACK SEAT IN JELLABY'S CART.
He said to her: "WILL YOU DO WHAT I WANT YOU TO?"
  She said to him: "THERE'S A SONG IN MY HEART".'

What was the Consequence? What did the World say?
  Hist, in the silence, to Damocles' sword!
Today Mr. Pomfret has left for Karachi
  And little Miss Montague screams in her ward.

HENRY REED

1914–

213

Chard Whitlow

(*Mr. Eliot's Sunday Evening Postscript*)

As we get older we do not get any younger.
Seasons return, and today I am fifty-five,
And this time last year I was fifty-four,
And this time next year I shall be sixty-two.
And I cannot say I should care (to speak for myself)
To see my time over again—if you can call it time,
Fidgeting uneasily under a draughty stair,
Or counting sleepless nights in the crowded Tube.

There are certain precautions—though none of them
  very reliable—
Against the blast from bombs, or the flying splinter,
But not against the blast from Heaven, *vento dei venti*,
The wind within a wind, unable to speak for wind;
And the frigid burnings of purgatory will not be touched
By any emollient.
                    I think you will find this put,
Better than I could ever hope to express it,
In the words of Kharma: 'It is, we believe,
Idle to hope that the simple stirrup-pump
Can extinguish hell.'
                    Oh, listeners,
And you especially who have turned off the wireless,
And sit in Stoke or Basingstoke, listening appreciatively
  to the silence,
(Which is also the silence of hell) pray, not for yourselves
  but your souls.

And pray for me also under the draughty stair.
As we get older we do not get any younger.

And pray for Kharma under the holy mountain.

R. P. LISTER

1914–

214 *Lament of an Idle Demon*

It's quiet in Hell just now, it's very tame,
  The devils and the damned alike lie snoring.
Just a faint smell of sulphur, not much flame;
  The human souls come here and find it boring.

Satan, the poor old Puritan, sits there
  Emitting mocking laughter once a minute;
Idly he scans a page of Baudelaire
  And wonders how he once saw evil in it.

He sips his brimstone at the Demons' Club
  (His one amusement now he's superseded)
And keeps complaining to Beelzebub
  That men make hotter hells than ever he did.

215 *The Revolutionaries*

O tremble, all ye earthly Princes,
  Bow down the crowned and chrism'd nob;
Wise is the Potentate that winces
  At the just clamour of the mob.

Shiver, ye Bishops, doff your mitres,
  Huddle between your empty pews
Here comes a horde of left-wing writers
  Brandishing salmon-pink reviews.

Comes the New Age. Your outworn faces
  Vanish at our enlightened curse,
While we erect in your old places
  Something considerably worse.

GAVIN EWART

1916–

216

Miss Twye

MISS TWYE was soaping her breasts in the bath
When she heard behind her a meaning laugh
And to her amazement she discovered
A wicked man in the bathroom cupboard.

VICTOR GRAY

1917–

217

Limericks

(i)

CHARLOTTE BRONTË said, 'Wow, sister! *What* a man!
He laid me face down on the ottoman:
  Now don't you and Emily
  Go telling the femily—
But he smacked me upon my bare bottom, Anne!'

(ii)

WHEN Gauguin was visiting Fiji
He said, 'Things are different here, e.g.
  While Tahitian skin
  Calls for tan, spread out thin,
You must slosh it on here with a squeegee.'

(iii)

WHEN our dean took a pious young spinster
On his cultural tour of York Minster,
  What they did in the clerestory
  Is rather a queer story—
But none of us hold it aginst her.

(iv)

A TAXI-CAB whore out at Iver
Would do the round trip for a fiver
  —Quite reasonable, too,
  For a sightsee, a screw,
And a ten-shilling tip to the driver.

(v)

THERE was a young fellow called Crouch
Who was courting a girl on a couch.
  She said, 'Why not a sofa?'
  And he exclaimed, 'Oh, for
Christ's sake shut your trap while I—ouch!'

(vi)

A YOUNG engine-driver called Hunt
Once took out his engine to shunt,
  Saw a runaway truck,
  And by shouting out, 'Duck!'
Saved the life of the fellow in front.

(vii)

ONE morning old Wilfrid Scawen Blunt
Was wanting a trip in a punt;
  But the puntmen had struck,
  So he shouted 'Good luck!
—Your wage is a social affront!'

(viii)

WHILE visiting Arundel Castle
I sent my sick uncle a parcel.
The contents of it
Were the local grey grit
To rub on his sore metatarsal.

(ix)

AN old East End worker called Jock
Lived a life full of danger and shock.
Even now, if one calls,
He will tell of his falls
In the Royal Victoria Dock.

CHARLES CAUSLEY

1917–

218

Betjeman, 1984

I SAW him in the Airstrip Gardens
(Fahrenheit at 451)
Feeding automative orchids
With a little plastic bun,
While above his brickwork cranium
Burned the trapped and troubled sun.

'Where is Piper? Where is Pontefract?
(Devil take my boiling pate!)
Where is Pam? and where's Myfanwy?
Don't remind me of the date!
Can it be that I am *really*
Knocking on for 78?

'In my splendid State Apartment
Underneath a secret lock
Finger now forbidden treasures
(Pray for me St. Enodoc!):
TV plate and concrete lamp-post
And a single nylon sock.

'Take your ease, pale-haired admirer,
  As I, half the century saner,
Pour a vintage Mazawattee
  Through the Marks and Spencer strainer
In a *genuine* British Railways
  (Luton Made) cardboard container.

'Though they say my verse-compulsion
  Lacks an interstellar drive,
Reading Beverley and Daphne
  Keeps *my* sense of words alive.
Lord, but *how* much beauty was there
  Back in 1955!'

TED PAUKER

1917–

219 *A Grouchy Good Night to the Academic Year*

(*with acknowledgments to W.M.P.*)

GOOD night to the Year Academic,
It finally crept to a close:
Dry fact about physic and chemic,
Wet drip about people and prose.
Emotion was down to a snivel
And reason was pulped to a pap,
Sociologists droning out drivel
And critics all croaking out crap.
For any such doctrine is preachable
In our tolerant Temple of Thought
Where lads that are largely unteachable
Learn subjects that cannot to be taught.

Good night to the Session—portentous
Inside the Vice-Chancellor's gown,
The personage who'll represent us
To Public and Party and Crown.
By enthusing for nitwitted novelty
He wheedles the moment'ry Great,
And at influence-dinner or grovel-tea
Further worsens the whims of the State.
So it is that, however much *we* rage,
The glibber of heart and of tongue
Build ladders to reach a life-peerage
From the buzz-sawed-up brains of the young.

Good night to the Session—the Chaplain,
Progressive and Ritualist too,
Who refers to the role of the apple in
Eden as 'under review'.
When the whole situation has ripened
Of his temporal hopes these are chief:
A notable increase in stipend,
And the right to abandon belief.
Meanwhile, his sermons: 'The Wafer—
Is it really the Presence of God?'
'Is the Pill or the French Letter Safer?'
And, 'Does the Biretta look Mod?'

Good night to the Session—what Art meant,
Or Science, no longer seemed plain,
But our new Education Department
Confuses confusion again.
'Those *teach* who can't *do*' runs the dictum,
But for some even that's out of reach:
They can't even teach—so they've picked 'em
To teach other people to teach.
Then alas for the next generation,
For the pots fairly crackle with thorn.
Where psychology meets education
A terrible bullshit is born.

Good night to the Session—the students
So eager to put us all right,
Whose conceit might have taken a few dents
But that ploughing's no longer polite;
So the essays drop round us in torrents
Of jargon a mouldering mound,
All worrying weakly at Lawrence,
All drearily pounding at Pound;
And their knowledge would get them through no test
On Ghana or Greece or Vietnam,
But they've mugged up enough for a Protest
—An easyish form of exam.

Good night to the Session—so solemn,
'Truth' and 'Freedom' their crusader crests,
One hardly knows quite what to call 'em
These children with beards or with breasts.
When from State or parental Golcondas
Treasure trickles to such little boys
They spend it on reefers and Hondas
—That is, upon sweeties and toys;
While girls of delicious proportions
Are thronging the Clinic's front stair,
Some of them seeking abortions
And some a psychiatrist's care.

Good night to the Session—the politics,
So noisy, and nagging, and null.
You can tell how the time-bomb of Folly ticks
By applying your ear to their skull;
Of course, that is only a metaphor,
But they have their metaphors too,
Such as 'Fascist', that's hardly the better for
Being used of a liberal and Jew
—The Prof. of Applied Aeronautics,
For failing such students as try,
With LSD lapping their cortex,
To fub up a fresh way to fly.

TED PAUKER

Good night to the Session—the Union:
The speeches with epigram packed,
So high upon phatic communion,
So low upon logic and fact.
(Those epigrams?—Oh well, at any rate
By now we're all quite reconciled
To a version that's vastly degenerate
From the Greek, via Voltaire and Wilde.)
Then the bold resolutions devoted
To the praise of a party or state
In *this* context most obviously noted
For its zeal in destroying debate.

Good night to the Session—the sculpture:
A jelly containing a clock;
Where they say, 'From the way that you gulped you're
Therapeutically thrilled by the shock!'
—It's the Shock of, alas, Recognition
At what's yearly presented as new
Since first seen at Duchamps' exhibition
'Des Maudits', in Nineteen-O-Two.
But let's go along to the Happening,
Where an artist can really unwind,
Stuff like 'Rapists should not take the rap' penning
In gamboge on a model's behind.

Good night to the Session—a later
Will come—and the freshmen we'll get!
Their pretensions will be even greater,
Their qualifications worse yet.
—But don't be too deeply depressible
At obtuseness aflame for applause;
The louts that are loudest in decibel
Melt away in post-graduate thaws.
Don't succumb to an anger unreasoned!
Most students are charming, and bright;
And even some dons are quite decent . . .
But good night to the Session, good night!

220

A Trifle for Trafalgar Day

(*with acknowledgments to G. K. C.*)

'Drake . . . Cabot . . . challenge . . . opportunity . . . courage . . .'—*The Prime Minister.*

WHO's the Dover-based day tripper
Heir to, Heath?
—The captain of the close-rigged clipper?
Is he, Heath?
(While for plump young brokers preening
On the day that Britain joins
'Rigging' has another meaning
And a 'clipper' is for coins . . .)
Was the cry of Drake and Raleigh,
Tacked into the tempest's teeth,
'Ah, thank God, at last—there's Calais!'
Was it, Heath?

Belgian bureaucrats in boardrooms
Differ, Heath,
From midshipmites in Nelson's wardrooms,
Don't they, Heath?
But would the *Victory* quite suit your
Grand adventures where they range,
When the Wave is of the Future
And the Winds are those of Change?
Though Industry is advantageous
Ask Sir Alec, ask Sir Keith,
Are its Captains so Courageous?
Are they, Heath?

You'll be needing loyal comrades,
Won't you, Heath?
—It's years since Germans came on bomb-raids,
Ain't it, Heath?
The red fool-fury of the Seine now
Has not raged since Sixty-eight.

In Rome for one whole British reign now
  Mere millions cheer the Total State
While in the countries of our cousins,
  Bowed our lousy laws beneath,
Revolutions come in dozens,
          Don't they, Heath?

Nowadays what does 'The Horn'
          Conjure, Heath?
Lorry-loads of Danish porn,
          Trader Heath?
Are economic blizzards raving
  Where the frozen assets clot
Conditions that you'll be braving
  Just like Frobisher or Scott . . .?
No, talk of profit and debenture
  Fix Liège with loans from Leith,
But the Spirit of Adventure . . .
          Chuck it, Heath!

221 *Garland for a Propagandist*

(*Air: The Vicar of Bray*)

IN good old Stalin's early days
When terror little harm meant
A zealous commissar I was
And so I got preferment.
I grabbed each peasant and I said
'Can there be something *you* lack?'
And if he dared to answer 'bread'
I shot him for a kulak.
    For on this rule I will insist
    Because I have the knack, Sir:
    Whichever way its line may twist
    I'll be a Party hack, Sir!

Then Stalin took the Secret Police
And gave it to Yagoda.
Many a Party pulse might cease
But I stayed in good odour.
At all the cases that he brought
I welcomed each confession,
And when he too turned up in court
I attended every session.

When Yezhov took the vacant place
And blood poured out in gallons
Thousands fell in dark disgrace
But I still kept my balance.
I studied, as the Chekists pounced,
The best way to survival
And almost every day denounced
A colleague or a rival.

When Yezhov got it in the neck
(In highly literal fashion)
Beria came at Stalin's beck
To lay a lesser lash on;
I swore our labour camps were few,
And places folk grew fat in;
I guessed that Trotsky died of flu
And colic raged at Katyn.

And when things once again grew hot
From Western war-psychosis,
I damned the 'cosmopolitan' lot
Because of their hook noses.
The Doctors should be shot, I cursed,
As filthy spy-recruiters.
But Stalin chanced to kick off first
—So I cursed their persecutors.

Malenkov, now our Party's head,
Tried out a tack quite new, Sir,
Saying what had never been said
—And so I said it too, Sir:
I boldly cried that clobber and scoff
Should go to the consumer
—But his overthrow soon tipped me off
This was a Right-wing bloomer.

When Khrushchev next came boldly on
Denouncing Stalin's terror,
I saw that what we'd so far done
Had mostly been an error.
My rivals all lay falsely framed
Under the Russian humus
And their innocence I now proclaimed
—Because it was posthùmous.

But Khrushchev guessed his chances wrong
And the present lot took over.
And I saw that though we'd suffered long
At last we were in clover;
Now Stalin's name I freely blessed,
A bonny, bonny fighter.
—And I told the intellectual West
When it's right to jug a writer.

Now the Collective Leadership
Of Brezhnev and Kosygin
I'll back until some rivals slip
By intricate intrigue in;
And, if the worst comes to the worst
And they're scragged in the Lubyanka,
I'll see they get as foully cursed
As any Wall Street banker.
    And on this rule I will insist
    Because I have the knack, Sir!
    Whichever way its line may twist
    I'll be a Party hack, Sir!

222

Limeraiku

THERE'S a vile old man
Of Japan who roars at whores:
'Where's your bloody fan?'

ROGER WODDIS

1917–

223 *Ethics for Everyman*

THROWING a bomb is bad,
Dropping a bomb is good;
Terror, no need to add,
Depends on who's wearing the hood.

Kangaroo courts are wrong,
Specialist courts are right;
Discipline by the strong
Is fair if your collar is white.

Company output 'soars',
Wages, of course, 'explode';
Profits deserve applause,
Pay-claims, the criminal code.

Daily the Church declares
Betting-shops are a curse;
Gambling with stocks and shares
Enlarges the national purse.

Workers are absentees,
Businessmen relax,
Different as chalk and cheese;
   Social morality
   Has a duality—
One for each side of the tracks.

224 *Nothing Sacred*

FROM his library in Surrey,
   Hung with regimental swords,
Came a cry as Colonel Curry
   Read the latest news from Lord's.

Turning to a bright vermilion,
  He could not believe his eyes:
'Jackets off in the pavilion.'
  (Members, though, must still wear ties.)

Never looked two cheeks more hollow,
  Nor less stiff an upper lip;
What in heaven's name could follow
  Gentlemen allowed to strip?

Is this, thought the baffled colonel,
  Why we fought the Second War?
Pages from his favourite journal
  Slithered to the parquet floor.

Guardsmen swelter in the city,
  Gallant typists still survive,
Yet the M.C.C. Committee
  Gets run out at 95.

Decent standards melt like butter,
  Cricket is becoming crude;
Ah, my friend, and oh! Calcutta!
  Soon they'll play it in the nude!

JOHN HEATH-STUBBS

1918–

225 FROM *An Ecclesiastical Chronicle*

THE year of Our Lord two thousand one hundred and seven,
The first electronic computer
Was appointed to a bishopric in the Church of England.
The consecration took place
At a Pontifical High Mass
In the new Cathedral of Stevenage,
In the presence of the Most Reverend
Mother in God, Her Grace Rita,
By divine Connivance *Cantuar. Archepiscopissa.*

Monsignor PFF-pff (75321/666)
With notable efficiency, tact, and benevolence, presided
For the next three hundred years
Over his diocese. (He had previously worked
In the mission field—rural Dean of Callisto,
One of the Satellites of Jupiter.)
After which he was honourably retired,
Only a little rusted, to the Science Museum
In South Kensington—there frequented and loved
By generations of schoolchildren.

As *The Times* remarked on that occasion,
'He stood for the best in the Anglican tradition':
In dubitable succession, one might say,
From our contemporary Dr.———, of ———.

226 *The Poet of Bray*

BACK in the dear old thirties' days
  When politics was passion
A harmless left-wing bard was I
  And so I grew in fashion:
Although I never really *joined*
  The Party of the Masses
I was most awfully chummy with
  The Proletarian classes.
    This is the course I'll always steer
      Until the stars grow dim, sir—
    That howsoever taste may veer
      I'll be in the swim, sir.

But as the tide of war swept on
  I turned Apocalyptic:
With symbol, myth and archetype
  My verse grew crammed and cryptic:
With New Romantic zeal I swore
  That Auden was a fake, sir,
And found the mind of Nicky Moore
  More int'resting than Blake, sir.

White Horsemen down New Roads had run
  But taste required improvement:
I turned to greet the rising sun
  And so I joined the Movement!
Glittering and ambiguous
  In villanelles I sported:
With Dr. Leavis I concurred,
  And when he sneezed I snorted.

But seeing that even John Wax might wane
  I left that one-way street, sir;
I modified my style again,
  And now I am a Beat, sir:
So very beat, my soul is beat
  Into a formless jelly:
I set my verses now to jazz
  And read them on the telly.

Perpetual non-conformist I—
  And that's the way I'm staying—
The angriest young man alive
  (Although my hair is greying)
And in my rage I'll not relent—
  No, not one single minute—
Against the base Establishment
  (Until, of course, I'm in it).
      This is the course I'll always steer
        Until the stars grow dim, sir—
      That howsoever taste may veer
        I'll be in the swim, sir.

GERRY HAMILL

1919–

227 *A Song of the GPO*

I'M the bloke that's trained to sit behind the public stamp machines
When you come to post a letter in the rain.
  'Ow I laugh to 'ear the curses
  As they fiddle in their purses
For a 10p piece that won't pop out again.

It's me job to put the rolls of stamps behind the little slot
So you get one when you pokes your money through.
  'Ow I giggle at the slangin'
  And the nasty-tempered bangin'
If it don't come out when it's supposed to do.

If the stamp machines gets busy I put up me 'empty' signs,
Then I makes the tea and 'as me little snack,
  But the stream of filthy language
  Doesn't put me off me sangwidge
'Cos I'm taught to smile and *never* answer back.

Now, the *proper* way to buy a stamp is from the counter clerk,
Who provides a queue where you can 'ang about;
  If you don't know any better
  Than to write yer flippin' letter
After five, then you deserve to go without.

PETER VEALE

1919–

228 *Bold Troubleshooters*

WE'RE 'er Majesty's bold troubleshooters; wherever they send us we goes.
Our job is to teach folks a lesson when they treads on 'er Majesty's toes.
If we don't rightly know what we're sent for—well, it's 'ardly for us to inquire,
An' we knows that the CO won't tell us; 'e just says the brush is afire.
We may 'ave to salvage a sheikdom or get a few niggers in line.
There's an airyplane waitin' to take us, so it's fly out an' clobber the swine!
We goes an' we does what we're told to, without any frettin' or fuss,
An' we can't ask the Queen why she sends us 'cause the Queen don't know no more than us.
(*Chorus*)
Oh the airyplane's waitin' at Lyneham an' our pickshers is all in the Press,
For 'er Majesty's bold troubleshooters is flyin' to sort out the mess.

Oh, Cyprus is just round the corner an' Aden's a ride on the bus;
An' we likes to drop in on Malaysher when the Bung's boys is causin' a fuss.
We steams and we stews in the jungle an' out in the desert we're browned;
We ain't 'ardly got any empire but by Gawd we don't 'arf get around!
Now they says we may go to Rhodeesher to put down a feller named Smith;
It seems 'e's some sort of a rebel, but 'e's white an' 'e says 'e's our kith.
We can't 'ave no shootin' or bloodshed, so Gawd knows 'ow Smith's goin' to crack,
An' we'd all be a ruddy sight 'appier if them rebel Rhodeeshuns was black.
(*Chorus*)
Oh the airyplane's, etc.

EDWARD BLISHEN

1920–

229

Abroad Thoughts

OH, not to be in England
Now that April's there!
R.B., you looked at England
Through a rosy pair
Of expatriate's goggles,
Basking by the Med.
Imagination boggles
At the lies you spread!
You're right: the leaves are tiny;
But you quite forgot
(Italy's sunshiny)
That the rain is not!
Oh, you Apriliser!
So the thrush sings prettily!
Wise? But you were wiser,
Robert, warm in Italy!
Let your pear tree scatter
Blossom on the clover:
You were in the latter
Five hundred miles from Dover!

D. J. ENRIGHT

1920–

230 *An Underdeveloped Country*

AND there are times truly
(This is no time for irony)
One can only surmise
That why the whole place isn't bugged
Is the initial cost of the enterprise.

It cannot be because it takes so many
Men to watch so many,
For unemployment's on the up and up
And listening keeps a lot of people quiet.

But one of these days surely
Some big nation
Will stake us to modernization.
We'll end with Russian mikes maybe
And U.S. tape-recorders. We
Are why and where the powers collaborate,
To teach us grown-up ways.

For there are times truly
(Which is no time for irony)
One can't be certain that our friends
And colleagues will denounce us.
They fail to, fairly often.
It seems that we are children
Still, we need that adult help
If we are ever to develop.

231 *Royalties*

As 'Name of individual, partnership, or corporation to whom paid'
I find my own, followed by (in brackets) 'Faust'.
The amount of income received in this capacity is $3.30 gross,
From which I am glad to see no tax has been withheld.

Egotist as he is
One had never thought the Devil so close-fisted.
He wasn't always:
Gretchen—Helen—contributions to knowledge—all that real estate . . .
What can have changed him?

And yet, Goethe only lasted a couple of months after completing his masterpiece,
So I could even be said to be lucky, nearly twenty-four years after publication, still making $3.30 out of a little crib of the Master's epic designed for non-German-speaking near-dropouts taking the World Literature course.

None of us, it seems, even though no tax is deducted, gets much for selling his soul.
The sea took back the land, Gretchen lost her head, Helen was incorporeal, the scholarship soon discredited;
Faust died the moment he started to enjoy life; and Goethe's poetry is supplanted by a crib—
It was always a buyer's market, always.

232 *Buy One Now*

THIS is a new sort of Poem,
It is Biological.
It contains a special Ingredient
(Pat. pend.) which makes it different
From other brands of poem on the market.

This new Poem does the work for you.
Just drop your mind into it
And leave it to soak
While you relax with the telly
Or go out to the pub
Or (if that is what you like)
You read a book.

It does the work for you
While (if that is what you like)
You sleep. For it is Biological
(Pat. pend.), it penetrates
Into the darkest recesses,
It removes the understains
Which it is difficult for us
Even to speak of.

Its action is so gentle
That the most delicate mind is unharmed.
This new sort of Poem
Contains an exclusive new Ingredient
(Known only to every jackass in the trade)
And can be found in practically any magazine
You care to mention.

VERNON SCANNELL

1922–

233

Poetry Reading

BEFORE the thing begins we have
Light conversation among those who know
Each other; and those who have come alone
Try to stick their profiles on your vision
Like stamps or children's transfers:
Others sit there looking sad and spiritual
And rather dreamy.
There are more women here than men.

Just before the poems are read
The atmosphere becomes subdued,
Silent and reverent like a chapel
Or a day of remembrance:
Then come the words, skilfully mouthed,
And the heads of the audience nod approval,
And eyes glow yet more spiritual.
They are really rather bored.

Soon their eyes begin to scan the room
And the poetry lullabies on like any sermon;
Then irreligious speculations intrude:
How interesting that man below the window looks;
Beards are really rather fascinating,
On certain types of course;
He looks so sensitive, yet cruel—
But we must concentrate on poetry.

Poetry is concentrated on:
The words come fluttering into focus
Announcing that the nightingales and noble sentiments
Have already been displayed, are now
Packed safely back into the conjurer's bag;
Now is the ripest moment, now
For the best trick in the pack, the one
That never fails to ring a bell.

Here it comes: the poets in a naughty mood!
The gods, unbending, let their trousers down,
And oh, the giggles and the wriggles as the words
Assume the sweet white wicked shapes
Of breasts, of bums, and things one dare not name.
The atmosphere has changed as if
The vicar had gone mad and danced upon
The altar table, vine leaves on his pate.

Lord Rochester, diluted but still hot,
Donne's log book of his explorations,
Adventures in the hot dark thickets
Of tremulous Americas;
And later Billy Blake and bad Lord Byron,
Brave modern poets who just don't give a damn—
The atmosphere grows thrilling prickles, but alas,
There are more women here than men.

KINGSLEY AMIS

1922–

234 *The Helbatrawss*

QVITE horfen, fer a lark, coves on a ship
  Ketches a uge sea-bird, a helbatrawss,
A hidle sod as mucks in on the trip
  By follerin the wessel on its course.

Theyve ardly got im on the deck afore,
  Cackanded, proper chokker—never mind
Es a igh-flier—cor, e makes em roar
  Voddlin abaht, is vings trailin beind.

Up top, yus, e was smashin, but es grim
  Like this: e aint alf hugly nah es dahned:
Vun perisher blows Voodbine-smoke at im,
  Anuvver tikes im orff by oppin rahnd!

A long-aired blokes the sime: ead in the clahds,
  E larfs at harrers, soups is cupper tea;
But dahn to earf in these ere bleedin crahds,
  Them uge great vings balls up his plates, yer see.

235 *After Goliath*

What shall be done to the man
that killeth this Philistine?
I Sam. xvii 26

THE first shot out of that sling
Was enough to finish the thing:
The champion laid out cold
Before half the programmes were sold.
And then, what howls of dismay
From his fans in their dense array:
From aldermen, adjutants, aunts,
Administrators of grants,

Assurance-men, auctioneers,
Advisers about careers,
And advertisers, of course,
Plus the obvious b——s in force:
The whole reprehensible throng
Ten times an alphabet strong.
But such an auspicious debut
Was a little too good to be true,
Our victor sensed; the applause
From those who supported his cause
Sounded shrill and excessive now,
And who were they, anyhow?
Academics, actors who lecture,
Apostles of architecture,
Ancient-gods-of-the-abdomen men,
Angst-pushers, adherents of Zen,
Alastors, Austenites, A-test
Abolishers—even the straightest
Of issues looks pretty oblique
When a movement turns into a clique,
The conqueror mused, as he stopped
By the sword his opponent had dropped:
Trophy, or means of attack
On the rapturous crowd at his back?
He shrugged and left it, resigned
To a new battle, fought in the mind,
For faith that his quarrel was just,
That the right man lay in the dust.

236 FROM *The Evans Country*

(i)

Aberdarcy: the main square

BY the new Boots, a tool-chest with flagpoles
Glued on, and flanges, and a dirty great
Baronial doorway, and things like port-holes,
Evans met Mrs. Rhys on their first date.

Beau Nash House, that sells Clothes for Gentlemen,
Jacobethan, every beam nailed on tight—
Real wood, though, mind you—was in full view when,
Lunching at the Three Lamps, she said all right.

And he dropped her beside the grimy hunk
Of castle, that with luck might one day fall
On to the *Evening Post*, the time they slunk
Back from that lousy week-end in Porthcawl.

The journal of some bunch of architects
Named this the worst town centre they could find;
But how disparage what so well reflects
Permanent tendencies of heart and mind?

All love demands a witness: something 'there'
Which it yet makes part of itself. These two
Might find Carlton House Terrace, St. Mark's Square,
A bit on the grand side. What about you?

(ii)

Langwell

'Now then, what are you up to, Dai?'
'Having a little bonfire, pet.'
  Bowed down under a sack,
With steps deliberate and sly,
His deacon's face full of regret,
  Evans went out the back.

Where no bugger could overlook
He dumped into a blackened bin
  Sheaves of photogravure,
Now and then an ill-printed book,
Letters in female hands: the thin
  Detritus of amour.

Paraffin-heightened flames made ash
Of *Lorraine Burnet in 3-D*
  And *I'm counting the days*
And *the head girl took off her sash*
And *Naturist* and *can we be*
  Together for always?

He piped an eye—only the smoke—
Then left that cooling hecatomb
  And dashed up to his den,
Where the real hot stuff was. A bloke
Can't give any old tripe house-room:
  Style's something else again.

(iii)

Pendydd

LOVE is like butter, Evans mused, and stuck
The last pat on his toast. Breakfast in bed
At the Red Dragon—when Miss Protheroe,
Wearing her weekday suit, had caught the train
Back home, or rather to her place of work,
United Mutual Trust—encouraged thought,
And so did the try-asking-me-then look
The bird who fetched the food had given him.
Scrub that for now. Love is like butter. It
Costs money but, fair play, not all that much,
However hard you go at it there's more,
Though to have nothing else would turn you up
(Like those two fellows on that raft,[1] was it?),
Nothing spreads thinner when you're running short;
Natural? Well, yes and no. Better than guns,
And—never mind what the heart experts say—
Let's face it, bloody good for you. Dead odd
That two things should turn out so much alike,
He thought, ringing the bell for more of both.

(iv)

Brynbwrla

LOVE's domain, supernal Zion,
  How thy rampart gleams with light,
Beacon to the wayworn pilgrim
  Stoutly faring through the night!

Some, their eyes on heavenly mansions,
  Tread the road their fathers trod,
Others, whom the Foe hath blinded,
  Far asunder stray from God.

And still others—take old Evans—
  Anchor on their jack instead;
Zion, pro- or non- or anti-,
  Never got them out of bed.

[1] Dinghy, actually. Evans is thinking of an episode in *The Bombard Story* (Penguin edition, p. 17).

Light's abode? There stands the chapel,
Flat and black against the sky.
Tall hotels ablaze with neon
Magnetise the sons of Dai.

(v)

Fforestfawr

WHEN they saw off Dai Evans's da
The whole thing was done very nice:
Bethesda was packed to the doors,
And the minister, Urien Price,
Addressed them with telling effect.

'Our brother grew rich in respect,'
He told them in accents of fire;
'A man of unshakable strength,
Whom to know was at once to admire.
He did nothing common or mean.'

They'd no notion of coming between
That poor young Dai and his grief,
So each of them just had a word
With him after, well-chosen and brief:
'I looked up to him, boy' sort of touch.

He thanked one and all very much,
But thought, as he waved them goodbye,
Was respect going to be what they felt
When Bethesda did honour to Dai?
No, something more personal, see?

'Hallo, pet. Alone? Good. It's me.
Ah now, who did you think it was?
Well, come down the Bush and find out.
You'll know me easy, because
I'm wearing a black tie, love.'

(vi)

Aldport (*Mystery Tour*)

HEARING how tourists, dazed with reverence,
Look through sunglasses at the Parthenon,
Dai thought of that cold night outside the Gents
When he touched Dilys up with his gloves on.

(vii)

Aberdarcy: the Chaucer road

5.40. The Bay View. After the office,
Evans drops in for a quick glass of stout,
Then, by the fruit-machine, runs into Haydn,
Who's marrying the kid he's nuts about.

Of course, he won't pretend it's all been easy:
The wife's three-quarters off her bloody head,
And Gwyneth being younger than their youngest
Leaves certain snags still to be combated.

Oh, no gainsaying that she's quite a handful;
No, not bad-tempered, man, just a bit wild.
He likes a girl to show a touch of spirit;
It's all the better when you're reconciled.

And then, dear dear, what dizzy peaks of passion!
Not only sex, but mind and spirit too,
Like in that thing Prof. Hughes took with the Honours:
That's right, *The Rainbow*—well, it's all come true.

6.10. The Humber. Evans starts reflecting
How much in life he's never going to know:
All it must mean to really love a woman.
He pulls up sharp outside a bungalow.

6.30. Balls to where. In like a whippet;
A fearsome thrash with Mrs. No-holds-barred
(Whose husband's in his surgery till 7);
Back at the wheel 6.50, breathing hard.

7.10. 'Braich-y-Pwll'.—'Hallo now, Megan.
No worse than usual, love. You been all right?
Well, this looks good. And there's a lot on later;
Don't think I'll bother with the club tonight.'

Nice bit of haddock with poached egg, Dundee cake,
Buckets of tea, then a light ale or two,
And 'Gun Smoke', 'Danger Man', the Late Night Movie—
Who's doing better, then? What about you?

PHILIP LARKIN

1922–

237

Fiction and the Reading Public

GIVE me a thrill, says the reader,
Give me a kick;
I don't care how you succeed, or
What subject you pick.
Choose something you know all about
That'll sound like real life:
Your childhood, your Dad pegging out,
How you sleep with your wife.

But that's not sufficient, unless
You make me feel good—
Whatever you're 'trying to express'
Let it be understood
That 'somehow' God plaits up the threads,
Makes 'all for the best',
That we may lie quiet in our beds
And not be 'depressed'.

For I call the tune in this racket:
I pay your screw,
Write reviews and the bull on the jacket—
So stop looking blue
And start serving up your sensations
Before it's too late;
Just please me for two generations—
You'll be 'truly great'.

238

Toads

WHY should I let the toad *work*
  Squat on my life?
Can't I use my wit as a pitchfork
  And drive the brute off?

Six days of the week it soils
With its sickening poison—
Just for paying a few bills!
That's out of proportion.

Lots of folk live on their wits:
Lecturers, lispers,
Losels, loblolly-men, louts—
They don't end as paupers;

Lots of folk live up lanes
With fires in a bucket,
Eat windfalls and tinned sardines—
They seem to like it.

Their nippers have got bare feet,
Their unspeakable wives
Are skinny as whippets—and yet
No one actually *starves*.

Ah, were I courageous enough
To shout *Stuff your pension!*
But I know, all too well, that's the stuff
That dreams are made on:

For something sufficiently toad-like
Squats in me, too;
Its hunkers are heavy as hard luck,
And cold as snow,

And will never allow me to blarney
My way to getting
The fame and the girl and the money
All at one sitting.

I don't say, one bodies the other
One's spiritual truth;
But I do say it's hard to lose either,
When you have both.

239 *Toads Revisited*

WALKING around in the park
Should feel better than work:
The lake, the sunshine,
The grass to lie on,

Blurred playground noises
Beyond black-stockinged nurses—
Not a bad place to be.
Yet it doesn't suit me,

Being one of the men
You meet of an afternoon:
Palsied old step-takers,
Hare-eyed clerks with the jitters,

Waxed-fleshed out-patients
Still vague from accidents,
And characters in long coats
Deep in the litter-baskets—

All dodging the toad work
By being stupid or weak.
Think of being them!
Hearing the hours chime,

Watching the bread delivered,
The sun by clouds covered,
The children going home;
Think of being them,

Turning over their failures
By some bed of lobelias,
Nowhere to go but indoors,
No friends but empty chairs—

No, give me my in-tray,
My loaf-haired secretary,
My shall-I-keep-the-call-in-Sir:
What else can I answer,

When the lights come on at four
At the end of another year?
Give me your arm, old toad;
Help me down Cemetery Road.

240 *I Remember, I Remember*

COMING up England by a different line
For once, early in the cold new year,
We stopped, and, watching men with number-plates
Sprint down the platform to familiar gates,
'Why, Coventry!' I exclaimed. 'I was born here.'

I leant far out, and squinnied for a sign
That this was still the town that had been 'mine'
So long, but found I wasn't even clear
Which side was which. From where those cycle-crates
Were standing, had we annually departed

For all those family hols? . . . A whistle went:
Things moved. I sat back, staring at my boots.
'Was that,' my friend smiled, 'where you "have your roots"?'
No, only where my childhood was unspent,
I wanted to retort, just where I started:

By now I've got the whole place clearly charted.
Our garden, first: where I did not invent
Blinding theologies of flower and fruits,
And wasn't spoken to by an old hat.
And here we have that splendid family

I never ran to when I got depressed,
The boys all biceps and the girls all chest,
Their comic Ford, their farm where I could be
'Really myself'. I'll show you, come to that,
The bracken where I never trembling sat,

Determined to go through with it; where she
Lay back, and 'all became a burning mist'.
And, in those offices, my doggerel
Was not set up in blunt ten-point, nor read
By a distinguished cousin of the mayor,

Who didn't call and tell my father *There
Before us, had we the gift to see ahead—*
'You look as if you wished the place in Hell,'
My friend said, 'judging from your face.' 'Oh well,
I suppose it's not the place's fault,' I said.

'Nothing, like something, happens anywhere.'

241 *Self's the Man*

OH, no one can deny
That Arnold is less selfish than I.
He married a woman to stop her getting away
Now she's there all day,

And the money he gets for wasting his life on work
She takes as her perk
To pay for the kiddies' clobber and the drier
And the electric fire,

And when he finishes supper
Planning to have a read at the evening paper
It's *Put a screw in this wall—*
He has no time at all,

With the nippers to wheel round the houses
And the hall to paint in his old trousers
And that letter to her mother
Saying *Won't you come for the summer.*

To compare his life and mine
Makes me feel a swine:
Oh, no one can deny
That Arnold is less selfish than I.

But wait, not so fast:
Is there such a contrast?
He was out for his own ends
Not just pleasing his friends;

And if it was such a mistake
He still did it for his own sake,
Playing his own game.
So he and I are the same,

Only I'm a better hand
At knowing what I can stand
Without them sending a van—
Or I suppose I can.

242 *A Study of Reading Habits*

WHEN getting my nose in a book
Cured most things short of school,
It was worth ruining my eyes
To know I could still keep cool,
And deal out the old right hook
To dirty dogs twice my size.

Later, with inch-thick specs,
Evil was just my lark:
Me and my cloak and fangs
Had ripping times in the dark.
The women I clubbed with sex!
I broke them up like meringues.

Don't read much now: the dude
Who lets the girl down before
The hero arrives, the chap
Who's yellow and keeps the store,
Seem far too familiar. Get stewed:
Books are a load of crap.

243 *Annus Mirabilis*

SEXUAL intercourse began
In nineteen sixty-three
(Which was rather late for me)—
Between the end of the *Chatterley* ban
And the Beatles' first LP.

Up till then there'd only been
A sort of bargaining,
A wrangle for a ring,
A shame that started at sixteen
And spread to everything.

Then all at once the quarrel sank:
Everyone felt the same,
And every life became
A brilliant breaking of the bank,
A quite unlosable game.

So life was never better than
In nineteen sixty three
(Though just too late for me)—
Between the end of the *Chatterley* ban
And the Beatles' first LP.

ANTHONY HECHT

1923–

244 *The Dover Bitch*

A Criticism of Life: for Andrews Wanning

SO there stood Matthew Arnold and this girl
With the cliffs of England crumbling away behind them,
And he said to her, 'Try to be true to me,
And I'll do the same for you, for things are bad
All over, etc., etc.'
Well now, I knew this girl. It's true she had read
Sophocles in a fairly good translation

And caught that bitter allusion to the sea,
But all the time he was talking she had in mind
The notion of what his whiskers would feel like
On the back of her neck. She told me later on
That after a while she got to looking out
At the lights across the channel, and really felt sad,
Thinking of all the wine and enormous beds
And blandishments in French and the perfumes.
And then she got really angry. To have been brought
All the way down from London, and then be addressed
As a sort of mournful cosmic last resort
Is really tough on a girl, and she was pretty.
Anyway, she watched him pace the room
And finger his watch-chain and seem to sweat a bit,
And then she said one or two unprintable things.
But you mustn't judge her by that. What I mean to say is,
She's really all right. I still see her once in a while
And she always treats me right. We have a drink
And I give her a good time, and perhaps it's a year
Before I see her again, but there she is,
Running to fat, but dependable as they come.
And sometimes I bring her a bottle of *Nuit d'Amour*.

ANTHONY BRODE

1923–

245 *Breakfast with Gerard Manley Hopkins*

'Delicious heart-of-the-corn, fresh-from-the-oven flakes are sparkled and spangled with sugar for a can't-be-resisted flavour.'—*Legend on a packet of breakfast cereal*

SERIOUS over my cereals I broke one breakfast my fast
With something-to-read-searching retinas retained by print on a packet;
Sprung rhythm sprang, and I found (the mind fact-mining at last)
An influence Father-Hopkins-fathered on the copy-writing racket.

Parenthesis-proud, bracket-bold, happiest with hyphens,
The writers stagger intoxicated by terms, adjective-unsteadied—
Describing in graceless phrases fizzling like soda siphons
All things, crisp, crunchy, malted, tangy, sugared and shredded.

Far too, yes, too early we are urged to be purged, to savour
Salt, malt and phosphates in English twisted and torn,
As, sparkled and spangled with sugar for a can't-be-resisted flavour,
Come fresh-from-the-oven flakes direct from the heart of the corn.

DESMOND SKIRROW

1924–1976

246 *Ode on a Grecian Urn summarized*

GODS chase
Round vase.
What say?
What play?
Don't know.
Nice, though.

BRIAN W. ALDISS

1925–

247 *Tom Wedgwood Tells*

YES, in the summer of 1773
On Pendle Hill amid the daisies, I,
On Joseph Priestley's daughter doting,
Kissed her with a passion rare
When old Lavoisier wasn't lurking nigh—
Kissed her as the down was floating
On the phlogisticated air.

JAMES MICHIE

1927–

248

Arizona Nature Myth

Up in the heavenly saloon
Sheriff sun and rustler moon
Gamble, stuck in the sheriff's mouth
The fag end of an afternoon.

There in the bad town of the sky
Sheriff, nervy, wonders why
He's let himself wander so far West
On his own; he looks with a smoky eye

At the rustler opposite turning white,
Lays down a king for Law, sits tight
Bluffing. On it that crooked moon
Plays an ace and shoots for the light.

Spurs, badge and uniform red,
(It looks like blood, but he's shamming dead),
Down drops the marshal, and under cover
Crawls out dogwise, ducking his head.

But Law that don't get its man ain't Law.
Next day, faster on the draw,
Sheriff creeping up from the other side
Blazes his way in through the back door.

But moon's not there. He's ridden out on
A galloping phenomenon,
A wonder horse, quick as light.
Moon's left town. Moon's clean gone.

CHARLES TOMLINSON

1927–

249 *A Word in Edgeways*

TELL me about yourself they
say and you begin to
tell them about yourself and
that is just the way I
am is their reply: they play
it all back to you in another
key, their key, and then in mid-
narrative they pay you a
compliment as if to say what a good
listener you are I am
a good listener my stay
here has developed my faculty I will
say that for me I will not
say that every literate male in
America is a soliloquist, a
ventriloquist, a strategic
egotist, an inveterate
campaigner-explorer over and
back again on the terrain of himself—what I will
say is they are not un-
interesting: they are simply
unreciprocal and yes it was a
pleasure if not an unmitigated
pleasure and I yes I did enjoy our
conversation goodnightthankyou

MARTIN FAGG

1929–

250

Elegy on Thomas Hood

O SPARE a tear for poor Tom Hood,
Who, dazed by death, here lies;
His days abridged, he sighs across
The Bridge of Utmost Size.

His *penchant* was for punning rhymes
(Some lengthy, others—shorties);
But though his *forte* was his life,
He died within his forties.

The Muses cried: 'To you we give
The crown of rhymester's bay, Thos'.
Thos mused and thought that it might pay
To ladle out the pay-thos.

He spun the gold yet tangled yarn
Of sad Miss Kilmansegg;
And told how destiny contrived
To take her down a peg.

But now the weary toils of death
Have closed his rhyming toil,
And charged this very vital spark
To jump his mortal coil.

PETER PORTER

1929–

251

A Consumer's Report

THE name of the product I tested is *Life*,
I have completed the form you sent me
and understand that my answers are confidential.

I had it as a gift,
I didn't feel much while using it,
in fact I think I'd have liked to be more excited.
It seemed gentle on the hands
but left an embarrassing deposit behind.
It was not economical
and I have used much more than I thought
(I suppose I have about half left
but it's difficult to tell)—
although the instructions are fairly large
there are so many of them
I don't know which to follow, especially
as they seem to contradict each other.
I'm not sure such a thing
should be put in the way of children—
It's difficult to think of a purpose
for it. One of my friends says
it's just to keep its maker in a job.
Also the price is much too high.
Things are piling up so fast,
after all, the world got by
for a thousand million years
without this, do we need it now?
(Incidentally, please ask your man
to stop calling me 'the respondent',
I don't like the sound of it.)
There seems to be a lot of different labels,
sizes and colours should be uniform,
the shape is awkward, it's waterproof
but not heat resistant, it doesn't keep
yet it's very difficult to get rid of:
whenever they make it cheaper they seem
to put less in—if you say you don't
want it, then it's delivered anyway.
I'd agree it's a popular product,
it's got into the language; people
even say they're on the side of it.
Personally I think it's overdone,
a small thing people are ready
to behave badly about. I think
we should take it for granted. If its
experts are called philosophers or market
researchers or historians, we shouldn't

care. We are the consumers and the last
law makers. So finally, I'd buy it.
But the question of a 'best buy'
I'd like to leave until I get
the competitive product you said you'd send.

GEORGE MACBETH

1932–

252

The Political Orlando

I

Orlando in Chicago

LAST year, Orlando
went to Chicago
for the famous party conference.

There he met,
one after another,
three jolly American policemen.

The first one
hit him with a night-stick
across the foreskin.

The second one
spat a wad of half-chewed shag
all over the blue grass of Illinois.

The third one
just stood and fingered the butt of his pistol
like it was some kind of a
surrogate penis.

After that experience,
Orlando came home,
a sadder but a wiser man.

Next year,
Orlando will go to the moon.

2

Orlando Goes to the Moon

In every particular,
we are the imperialists of the Sea of Tranquillity
according to Orlando,

or will be,
when he gets there.

Three for the sea-saw:
the Plain of Jars
the Bay of Pigs
and the Sea of Tranquillity

Orlando knows them all,
their qualities,
and how to put quite a good face
on why the Stars and Stripes
had to be pinned there.

Quite a story,
how Orlando got his Purple Heart.

3

Orlando in Vietnam

Now it can be told:

Orlando is on
a search and destroy mission.

Over the roof-tops
he goes with his claws out
in a chopper.
From a small 7-transistor radio set
he will keep in earshot
of the top thirty.
Some of them, a select few,
may even be in there with him,
wide-eyed
and bushy-tailed.

Orlando is facing out
the VCs
eyeball to eyeball.

If he isn't careful,
Orlando is going to win
another medal.

Orlando will have a
ticker-tape welcome
back in Yawnsville, Massachusetts,

where the worst that ever happens
is the local dentist
having his wisdom teeth out.

Ah me, thought Orlando
(reading this)
if only a few others,
naming no names,
had had their wisdom teeth put in!

Orlando is quite a philosopher.

253 *The Orlando Commercial*

I

EEK!

Her legs are caught in something.

What appalling catastrophe
has trapped and is so
atrociously torturing
this beautiful naked girl's
legs?

Why, ORLANDO, of course.

Elusive, exclusive Orlando
  the new seamless nylon
has stolen up in secret
  and caught her in his lure.

2

Orlando yawns.

He is tired of being a fabric.

    He wants out.

Heh, you can't do that
  you'll tear the screen!

The screen tears.

Just shows.

It can't have been made
  of elusive, exclusive ORLANDO
the new seamless nylon.

254

The Five-Minute Orlando Macbeth

I

ACT I

  Orlando hails
the weird sisters
  and rides homes like a maniac.

  Eschewing the
rooky wood, he
  gallops across four blasted heaths
  towards his castle.

  There, washing her hands

  HERE'S YOUR TRAILER

Lady Orlando
stands
with harness on her back.

DUNC'S HERE
she says

HI, DUNC
he calls

2

Meanwhile, Lady O
screws his courage to the sticking-point
and they have a stiff night together.

Pretty probably.

Next, Orlando dips
the bloody dagger
in his wife's history book

and sets about gilding the faces of the grooms withal.

UNFORTUNATELY

it keeps cropping up
at awkward times
like
in his dreams, during banquets, etc.

3

MOREOVER

Hired lads
got in to put the
Macduff fry

out of their misery
don't help much,

though admittedly going in the catalogue for men.

4

Finally, it's

WITCH TIME AGAIN

and Thane Orlando
runs his eye along the cauldrons

BAD NEWS THERE, BUD
but
he takes it like a man
being, as he says,
so far advanced in gore
to return were as tedious as go o'er.

SO

5
ACT V
Drums. Trumpets. Marching trees

and the old woman with her bad scene
about all the perfumes of Arabia
not being a spot-lifter

RIGHT

not of woman born
untimely ripp'd

a hard choice, but
we all know it has to go at last to the goodies

THANK YOU, MAC ORLANDO

and stick his bloody head on the battlements
on your way out, please.

ALAN BENNETT

1934–

255

Place-Names of China

BOLDING Vedas! Shanks New Nisa!
Trusty Lichfield swirls it down
To filter beds on Ruislip Marshes
From my lav in Kentish Town.

The Burlington! The Rochester!
Oh those names of childhood loos—
Nursie rattling at the door-knob:
'Have you done your Number Twos?'

Lady typist—office party—
Golly! All that gassy beer!
Tripping home down Hendon Parkway
To her Improved Windermere.

Here I sit, alone and sixty,
Bald and fat and full of sin,
Cold the seat and loud the cistern
As I read the Harpic tin.

JOHN FULLER

1937–

256

Blues

YES, he's got her now,
She has her car and fur,
He's hot for her (and how)
But it's not any fun for her:
She has her fur (yes, sir)
But it's not any fun for her.

He's tip top now he's got
His cut, his fat *ten per*
(I'll bet he's had the lot)
But it's not any fun for her:
His fat *ten per* (yes, sir)
But it's not any fun for her.

He's far too fat, and who
But the ego can say *Prr*
For sun-tan oil and tea for two?
But it's not any fun for her:
She can say *Prr* (yes, sir)
But it's not any fun for her.

You see her and you buy,
She can say Yes, but sir,
One day she may ask why,
For it's not any fun for her:
She may say Yes (and how you try)
But it's not any fun for her.

SIMON CURTIS

1943–

257

Satie, at the End of Term

THE mind's eye aches from Henry James,
Like arms from heavy cases, lugged for miles.
Theme and structure, imagery and tone.

From Lawrence, too: how hard I dug
For insights sunk, yards deep, in turgid prose.
Theme and structure, imagery and tone.

Web of necessity in *Daniel Deronda*,
Gloom in *Dorrit*, gloom in Flaubert,
One more week to go, at
Theme and structure, imagery and tone.

SIMON CURTIS

So fitful-fresh as April sun,
You're welcome, clown;
Your good melodic dissonance
Will pierce low clouds of syllabus
  With humour's grace,
  Mercy of irreverence.

ACKNOWLEDGEMENTS

MY grateful thanks are due to the following for their kind assistance: Susan Allison, Jonathan Barker, Julian Barnes, Robert Conquest, Roy Fuller, Paul Fussell, Ian Hamilton, Anthony Hobson, Alexander Llewelyn, Douglas Matthews, Christopher Milne, Charles Monteith, John Julius Norwich, George Thomson, Anthony Thwaite and Antony Walker.

The publishers wish to thank the following, who have given them permission to reproduce copyright poems:

Brian W. Aldiss: 'Tom Wedgwood Tells' (© 1976 by Southmoor Serendipity). Originally published as a postcard by the Bellevue Press, Binghampton, New York, under the title 'Summer 1773'.

Kingsley Amis: Poems from 'The Evans Country' and 'After Goliath' from *A Look Round the Estate*. Reprinted by permission of Jonathan Cape Ltd, *Punch* and Harcourt Brace Jovanovich Inc. 'The Helbatrawss'. Reprinted by permission of *The Spectator*.

W. H. Auden: An extract from 'Letter to Lord Byron' from *Collected Longer Poems*. Reprinted by permission of Faber & Faber Ltd, and Curtis Brown Inc., New York. 'On the Circuit' from *About the House*, 'Doggerel by a Senior Citizen' from *Epistle to a Godson*. Reprinted by permission of Faber & Faber Ltd. All the above poems reprinted also from *Collected Poems* edited by Edward Mendelson (Copyright 1937, 1946, © 1965, 1969, by W. H. Auden) by permission of Random House Inc. 'Uncle Henry' (Copyright © 1966 by W. H. Auden), 'The Unknown Citizen' (Copyright 1940 and renewed 1968 by W. H. Auden), and 'Under Which Lyre' from *Collected Shorter Poems 1927–1957*. Reprinted by permission of Faber & Faber Ltd, and Random House Inc.

Max Beerbohm: 'A Luncheon' from *Max in Verse: Rhymes and Parodies by Max Beerbohm*, edited by J. G. Riewald. Reprinted by permission of William Heinemann Ltd, and The Stephen Green Press. 'Police Station Ditties' from *A Christmas Garland*. Reprinted by permission of Sir Rupert Hart-Davis on behalf of the Beerbohm Estate.

Hilaire Belloc: 'On a General Election', 'Fatigue' and 'Lord Finchley' from *Sonnets and Verse* (Duckworth). Reprinted by permission of A. D. Peters & Co. Ltd. 'The Yak', 'Extract from an ABC', and 'Rebecca Who Slammed Doors for Fun' from *Cautionary Verses*. Reprinted by permission of Gerald Duckworth & Co. Ltd., and Alfred Knopf, Inc.

E. C. Bentley: From *Clerihews Complete* (T. Werner Laurie Ltd). Reprinted by permission of A. P. Watt & Son on behalf of Nicolas Bentley.

ACKNOWLEDGEMENTS

John Betjeman: 'A Subaltern's Love-song', 'How to Get On in Society', 'In Westminster Abbey', 'Pot Pourri from a Surrey Garden', 'Invasion Exercise on the Poultry Farm' and 'Longfellow's Visit to Venice', from *Collected Poems.* Reprinted by permission of John Murray (Publishers) Ltd, and Houghton Mifflin Company. 'Executive' from *A Nip in the Air, Poems* (Copyright © 1974 by John Betjeman). Reprinted by permission of John Murray (Publishers) Ltd, and W. W. Norton & Co. Inc.

Edward Blishen: First printed in the *New Statesman* (17 April 1954). Reprinted by permission of The Statesman & Nation Publishing Co. Ltd.

Anthony Brode: Reprinted by permission of *Punch.*

Alan Brownjohn: Extract from 'Elizabeth Pender's Dream of Friendship' from *Warriors Career.* Reprinted by permission of Macmillan London and Basingstoke.

A. Butler: From *The Complete Limerick Book* by Lanford Reed (Jarrolds (Publishers) London Ltd). Reprinted by permission of Hutchinson Publishing Group Ltd.

Roy Campbell: From *Adamastor* (Faber & Faber Ltd). Reprinted by permission of Curtis Brown Ltd, on behalf of the Estate of Roy Campbell.

Charles Causley: From *Collected Poems 1951–1975* (Macmillan London and Basingstoke). Reprinted by permission of David Higham Associates Ltd.

G. K. Chesterton: 'Triolet' from *Coloured Lands* (Sheed & Ward, Ltd). Reprinted by permission of A. P. Watt & Son on behalf of Miss D. E. Collins. 'Variations on an Air', 'A Ballad of Abbreviations', 'Antichrist, or the Reunion of Christendom', and 'The Rolling English Road' from *The Collected Poems of G. K. Chesterton* (Methuen & Co. Ltd). (Copyright 1932 by Dodd, Mead & Company Inc., and renewed 1959 by Oliver Chesterton.) Reprinted by permission of A. P. Watt & Son on behalf of the Estate of the late G. K. Chesterton and Dodd, Mead and Co. Inc.

Noël Coward: From *The Lyrics of Noël Coward* (© Noël Coward 1965) (William Heinemann Ltd.). Reprinted by permission of Curtis Brown Ltd. on behalf of the Estate of Sir Noël Coward and the Overlook Press.

Simon Curtis: From *On the Abthorpe Road and Other Poems.* Reprinted by permission of Davis-Poynter Ltd.

Anthony C. Deane: 'An Ode', from *A Nonsense Anthology* edited by Carolyn Wells; 'The Cult of the Celtic', from *A Parody Anthology* edited by Carolyn Wells. Both reprinted by permission of Dover Publications Inc.

Paul Dehn: From *Fern on the Rock* (Copyright © 1965, 1976, by Dehn Enterprises Ltd). Reprinted by permission of Hamish Hamilton Ltd, London.

Walter de la Mare: From *The Complete Poems of Walter de la Mare 1969,* (Faber & Faber Ltd). Reprinted by permission of the Literary Trustees of Walter de la Mare and The Society of Authors as their representative.

Peter de Vries: 'Bacchanal' (Copyright 1960 by Peter de Vries), 'Christmas Family Reunion' (Copyright 1949 by Peter de Vries). From *The Tents of Wickedness.* Both poems originally appeared in *The New Yorker* and are reprinted by permission of Laurence Pollinger Ltd, and Little, Brown and Company.

T. S. Eliot: From '*Old Possum's Book of Practical Cats*' (Copyright 1939 by T. S. Eliot, Copyright 1967 by Esme Valerie Eliot). Reprinted by permission of Faber & Faber Ltd, and Harcourt Brace Jovanovich Inc.

Colin Ellis: From *Mournful Numbers*. Reprinted by permission of Macmillan London and Basingstoke.

D. J. Enright: 'Buy One Now', from *Sad Ires*, 'Royalties' and 'An Underdeveloped Country' from *Daughters of the Earth* (both Chatto & Windus Ltd). Reprinted by permission of Bolt & Watson Ltd.

Gavin Ewart: From *Poems and Songs* (The Fortune Press). Reprinted by permission of the author.

Martin Fagg: First printed in the *New Statesman* (27 August 1971). Reprinted by permission of The Statesman & Nation Publishing Co. Ltd.

Robert Frost: From *The Poetry of Robert Frost*, edited by Edward Connery Lathem (Copyright 1936 by Robert Frost, Copyright © 1964 by Lesley Frost Ballantine, Copyright © 1969 by Holt, Rinehart and Winston). Reprinted by permission of the Estate of Robert Frost, Jonathan Cape Ltd, and Holt, Rinehart and Winston.

John Fuller: First printed in the *New Statesman* (17 October 1969). Reprinted by permission of The Statesman & Nation Publishing Co. Ltd.

Kenneth Grahame: From *The Wind in the Willows* (Text Copyright University Chest Oxford). Reprinted by permission of Methuen's Children's Books and Curtis Brown Ltd, on behalf of Charles Scribner's Sons and the Estate of Kenneth Grahame.

Robert Graves: 'Wigs and Beards', 'Epitaph on an Unfortunate Artist', 'Wm. Brazier' and 'The Persian Version' from *Collected Poems*, '¡Wellcome, to the Caves of Artá!' from *Poems and Satires* (both Cassell & Co. Ltd). Reprinted by permission of A. P. Watt & Son on behalf of Mr. Robert Graves.

Gerry Hamill: First printed in the *New Statesman* (26 March 1976). Reprinted by permission of The Statesman & Nation Publishing Co. Ltd.

Thomas Hardy: From *The New Wessex Edition of the Complete Poems of Thomas Hardy*. Reprinted by permission of the Trustees of the Hardy Estate, The Macmillan Company of Canada, and Macmillan London and Basingstoke.

M. E. Hare: From *The Complete Limerick Book* by Langford Reed (Jarrolds (Publishers) London Ltd). Reprinted by permission of Hutchinson Publishing Group Ltd.

John Heath-Stubbs: From *Selected Poems* (Oxford University Press). Reprinted by permission of David Higham Associates Ltd.

Anthony Hecht: *The Hard Hours* (Copyright © 1958, 1960, 1961, 1966 by Anthony Hecht). Reprinted by permission of the author, Oxford University Press and Atheneum Publishers, Inc.

A. P. Herbert: From *Less Nonsense*. Reprinted by permission of A. P. Watt & Son on behalf of the Estate of the late Sir Alan Herbert and Methuen & Co. Ltd.

ACKNOWLEDGEMENTS

A. E. Housman: 'Occasional Poem' from *Collected Poems*. 'The Elephant' and 'Infant Innocence' from *A.E.H.* by Laurence Housman. Reprinted by permission of The Society of Authors as the Literary representative of the Estate of A. E. Housman, and Jonathan Cape Ltd, as publishers.

W. R. Inge: From *The Lure of the Limerick* edited by William S. Baring-Gould (Copyright © 1967 by William S. Baring-Gould. Reprinted by permission of Hart-Davis MacGibbon Ltd/Granada Publishing Ltd, Clarkson, N. Potter Inc. and A. P. Watt & Son on behalf of Mrs. L. Baring-Gould.

Hugh Kingsmill: From *The Best of Hugh Kingsmill*, edited by Michael Holroyd. Reprinted by permission of Victor Gollancz Ltd, and the McGraw-Hill Book Company.

Rudyard Kipling: From *The Definitive Edition of Rudyard Kipling's Verse*. Reprinted by permission of A. P. Watt & Son on behalf of The National Trust, Macmillan London and Basingstoke, and Doubleday & Company Inc.

E. V. Knox: From *Poems from Punch, 1909–1920*. Reprinted by permission of *Punch*.

R. A. Knox: From *The Complete Limerick Book* by Lanford Reed (Jarrolds (Publishers) London Ltd). Reprinted by permission of Hutchinson Publishing Group Ltd.

Osbert Lancaster: From *Façades and Faces*. Reprinted by permission of John Murray (Publishers) Ltd.

Philip Larkin: 'Self's the Man', 'A Study of Reading Habits' and 'Toads Revisited', from *The Whitsun Weddings* (Copyright © 1964 by Philip Larkin). Reprinted by permission of Faber & Faber Ltd. 'Annus Mirabilis' from *High Windows* (Copyright © 1974 by Philip Larkin). Reprinted by permission of Faber & Faber Ltd, and Farrar, Straus & Giroux Inc. 'Toads' and 'I Remember, I Remember' from *The Less Deceived*. Reprinted by permission of the Marvell Press. 'Fiction and the Reading Public' first appeared in *Essays in Criticism* and is reprinted by permission of the author.

C. S. Lewis: From *Poems*, edited by Walter Hooper (Copyright © 1964 by the Executors of the Estate of C. S. Lewis). Reprinted by permission of Collins Publishers, and Harcourt Brace Jovanovich Inc.

D. B. Wyndham Lewis: From *A Choice of Comic and Curious Verse*, edited by J. M. Cohen (Penguin Books Ltd). Reprinted by permission of A. D. Peters Ltd, on behalf of the Executor of D. B. Wyndham Lewis.

R. P. Lister: Reprinted by permission of *Punch*.

George Macbeth: From *The Orlando Poems* (Macmillan). Reprinted by permission of the author.

Phyllis McGinley: From *The Love Letters of Phyllis McGinley*. Reprinted by permission of Martin Secker & Warburg. Also from *Times Three* (Copyright 1954 by Phyllis McGinley). Originally appeared in *The New Yorker* and reprinted by permission of The Viking Press Inc.

Louis MacNeice: From *The Collected Poems of Louis MacNeice*, edited by E. R. Dodds (Copyright © 1966 by the Estate of Louis MacNeice).

ACKNOWLEDGEMENTS

Reprinted by permission of Faber & Faber Ltd, and Oxford University Press Inc.

James Michie: From *Possible Laughter* (Rupert Hart-Davis Ltd). Reprinted by permission of the author.

A. A. Milne: Extract from *Year In, Year Out* (Methuen & Co. Ltd). Reprinted by permission of Curtis Brown Ltd, London, on behalf of the Estate of A. A. Milne.

J. B. Morton: From *The Best of Beachcomber* (William Heinemann Ltd). Reprinted by permission of the author.

Ted Pauker: 'Limeraiku' first printed in the *New Statesman* (13 February 1976). Reprinted by permission of the author and The Statesman & Nation Publishing Co. Ltd. 'A Trifle for Trafalgar Day' first printed in *The Spectator* (28 October 1972) and reprinted with their permission and that of the author. 'Garland for a Propagandist' first printed in the *Times Literary Supplement*. Reprinted by permission of *The Times* and the author. 'A Grouchy Good Night to the Academic Year' first printed in the *Times Educational Supplement*. Reprinted by permission of the author.

J. R. Pope: Reprinted by permission of *Punch*.

Peter Porter: From *The Last of England* (© 1970 by Oxford University Press). Reprinted by permission of the author and Oxford University Press.

Henry Reed: From *A Map of Verona*. Reprinted by permission of Jonathan Cape Ltd.

Vernon Scannell: From *A Mortal Pitch* (Eyre & Spottiswoode (Publishers), Ltd). Reprinted by permission of the author.

Owen Seaman: 'England Expects' from *Salvage*. 'A Plea for Trigamy', 'The Sitting Bard' and part of 'A Birthday Ode to Mr. Alfred Austin' from *Owen Seaman: A Selection*. Reprinted by permission of Methuen and Co. Ltd.

Stanley J. Sharpless: First printed in the *New Statesman* (30 September 1977). Reprinted by permission of The Statesman & Nation Publishing Co. Ltd.

Desmond Skirrow: First printed in the *New Statesman* (30 July 1960). Reprinted by permission of The Statesman & Nation Publishing Co. Ltd.

J. C. Squire: From *Collected Parodies* (Hodder & Stoughton Ltd). Reprinted by permission of Mr. Raglan Squire.

L. A. G. Strong: From *The Body's Imperfections*. Reprinted by permission of Methuen & Co. Ltd.

Charles Tomlinson: From *The Way of a World*. Reprinted by permission of the author and Oxford University Press.

Peter Veale: First printed in the *New Statesman* (31 December 1965). Reprinted by permission of The Statesman & Nation Publishing Co. Ltd.

Roger Woddis: 'Ethics for Everyman' and 'Nothing Sacred' were first printed in the *New Statesman* (5 February 1971 and 2 July 1976). Reprinted by permission of The Statesman & Nation Publishing Co. Ltd.

P. G. Wodehouse: First printed in *Punch* (1901). Reprinted by permission of A. P. Watt & Son on behalf of the Estate of the Late P. G. Wodehouse.

ACKNOWLEDGEMENTS

'We're All Dry', 'The Sow Came in', 'Trip upon Trenchers', 'Rub-a-dub-dub', 'Tweedledum and Tweedledee', 'Baby, Baby, Naughty Baby', and 'I Saw a Fishpond' are taken from *The Oxford Dictionary of Nursery Rhymes*, edited by Iona and Peter Opie. Reprinted by permission of Oxford University Press.

'Dear Sir, Your astonishment's odd', 'There's a notable family named Stein', 'A vice most obscene and unsavoury', and 'An Argentine gaucho named Bruno' from *The Lure of the Limerick* edited by William S. Baring-Gould (Copyright © 1967 by William S. Baring-Gould). Reprinted by permission of Hart-Davis MacGibbon Ltd./Granada Publishing Ltd, Clarkson N. Potter Inc. and A. P. Watt & Son on behalf of Mrs. L. Baring-Gould.

The following poems are being printed for the first time in this anthology and appear by permission of their respective authors:

Alan Bennett: 'Place-Names of China' (© 1978 by Alan Bennett).
Victor Gray: 'Charlotte Brontë said', 'When Gauguin was visiting Fiji', 'When our dean took a pious young spinster', 'A taxi-cab whore out at Iver', 'There was a young fellow called Crouch', 'A young engine-driver called Hunt', 'One morning old Wilfrid Scawen Blunt', 'While visiting Arundel Castle' and 'An old East End worker called Jock' (© 1978 by Victor Gray).
Stuart Howard-Jones: 'Hibernia' (© 1978 by the Estate of the late Stuart Howard-Jones).
Anthony Powell: 'Caledonia' (© 1978 by Anthony Powell). Privately printed by Desmond Ryan, 1934.
Wynford Vaughan-Thomas: 'Hiraeth in N.W.3', 'To His Not-So-Coy Mistress' and 'Farewell to New Zealand' (© 1978 by L. J. W. Vaughan-Thomas).

While every effort has been made to secure permission, it has in a few cases proved impossible to trace the author or his executor. We apologize for our apparent negligence.

NOTES AND REFERENCES

THE order of the poems is chronological. Those of known authorship are arranged by the dates of their authors' births; no such precision is attainable with the anonymous pieces, which are assigned, wherever convenient in groups, to the likely period of their first circulation.

When necessary the poems have been modernized in spelling, punctuation and such matters as the use of italics and initial capitals, except in cases ('Hibernia', 'Caledonia') where archaic styles are intended for purposes of pastiche, or are an idiosyncrasy of the author.

No references in the Notes are given to poems by authors whose collected poems are easily available. For details of poems still in copyright, see also the Acknowledgements.

1. William Shakespeare. *The Tempest*, II, ii.
3. Anon. 'Oh that my lungs' in Carolyn Wells (ed.), *A Nonsense Anthology* (new edition 1958), where it is entitled 'Nonsense'. The fourth line of verse 1 there reads: 'That offer wary windmills to the rich'. Doubts have been expressed as to the authenticity of this poem.
4. Anon. 'Hye Nonny Nonny Noe' in John Aubrey, *Brief Lives* (1949), from the entry on John Overall.
5, 6. Samuel Butler. 5. R. Lamar (ed.), *Samuel Butler: Satires and Miscellaneous Poetry and Prose* (1928)—MSS. Butler and Thyor, British Museum (from 1885); 6. *Hudibras*, The First Part, Canto I, lines 119–234, Canto III, lines 1265–1310, The Second Part, Canto I, lines 123–74 and 343–78, The Third Part, Canto I, lines 1263–6, 1271–88, 1303–6.
8. Anon. 'Three Children' in Iona and Peter Opie (eds.), *The Oxford Dictionary of Nursery Rhymes* (1951).
9. Anon. Attributed variously to Rochester and to Milton.
10. Thomas Flatman. J. R. Tutin (ed.), *Poems and Songs* (1906), where it appears as Part II of 'A Bachelor's Song'.
14. Daniel Defoe. *The True-Born Englishman, A Satire* (1700), lines 56–69, 135–52, 175–94, 334–47, 358–67, 451–64, 479–82, 485–90, 499–500, 503–8.
15. Samuel Wesley. J. C. Squire (ed.), *Apes and Parrots* (n.d.).
16. William Walsh. *Poetical Miscellanies*, Vol. V. (1704).
18, 19. Jonathan Swift. Davis (ed.), *Poetical Works* (1967). 18. lines 1–90 only; 19. lines 73–146, 169–78, 225–52.
21. Anon. 'Drinking Song' in *Songs* (n.d., 19th century).
22. Anon. 'The Vicar of Bray' in *The British Musical Miscellany* (1734).

23. Anon. 'We're All Dry' in Iona and Peter Opie (eds.), *The Oxford Dictionary of Nursery Rhymes* (1951).
24–27. John Byrom. **24**, **26** and **27**. *Miscellaneous Poems* (1773); **24.** Some readings adopted from 2nd edition (1814); **25.** *Swift's Miscellanies* (1728).
28. Benjamin Franklin. T. H. Russell (ed.), *The Sayings of Poor Richard—Wit, Wisdom and Humor of Benjamin Franklin 1733–1758.*
29. Thomas Lisle. J. C. Squire (ed.), *The Comic Muse* (n.d.).
30. Samuel Johnson. 'Epigram' is taken from 'Burlesque of Lines by Lope de Vega'; 'Ballad' is from 'Parodies of "The Hermit of Warkworth" (11)'; 'Idyll' is from 'Parody of Thomas Warton'.
31, 32. Anon. 'The Sow Came In' and 'Trip Upon Trenchers' in Iona and Peter Opie (eds.), *The Oxford Dictionary of Nursery Rhymes* (1951).
35–37. Anon. **35.** *Christmas Box*, Vol ii (1798); **36.** J. Harris (ed.), *Original Ditties for the Nursery* (*c.* 1805); **37.** *Notes and Queries* (1877), printed in Iona and Peter Opie (eds.), *The Oxford Dictionary of Nursery Rhymes* (1951).
40. Ebenezer Elliott. Edwin Elliott (ed.), *The Poetical Works of Ebenezer Elliott*, (1876), where the title is 'Epigram'.
42. Thomas Love Peacock. *The Misfortunes of Elphin* (1829).
43. Anon. 'The Night Before Larry Was Stretched.' This version is a collation of W. H. Auden's text (*The Oxford Book of Light Verse* (1938)), that used in *Oliver's Book of Comic Songs* (n.d., 19th century), and personally collected variants.
44. Anon. 'In Peterborough Churchyard' in *Epigrams* (n.d., 19th century).
45–47. George Gordon Noel, Lord Byron. **45.** Verses 1–20 only. Sub-titled 'A Venetian Story'; **46.** Canto I xxxviii–xlv, Canto III xci–c, Canto XI xxix–xli, Canto XII lviii–lxiii, Canto XIV viii–xvii; **47.** from *Don Juan*, Canto I entitled 'Fragment'.
48. Hartley Coleridge. *Notes and Queries* (19 June 1869). Originally untitled.
49, 50. J. R. Planché. *Vauxhall Comic Song Book.* **50.** from 'London Exhibitions'.
51. Alaric A. Watts. *The Trifler* (Westminster School Magazine) (1817)
55, 56. Anon. 'Mr. and Mrs. Vite's Journey' and 'A Maiden There Lived' in *Oliver's Book of Comic Songs* (n.d., 19th century).
67, 68. W. M. Thackeray. **67.** Samuel Bevan, *Sand and Canvas; A Narrative of Adventures in Egypt with a Sojourn among Artists in Rome* (1849), where it was entitled 'The Three Sailors'; **68.** *Ballads and Verses and Miscellaneous Contributions to Punch* (1904).
72, 73. Shirley Brooks. *Wit and Humour* (1875).
75, 76. C. G. Leland. *The Breitmann Ballads* (1902). **75.** original title, 'De Maiden Mit Nodings On'.
77. Anon. 'I saw a Fishpond' in *Folk-Lore* (1889). Printed in Iona and Peter Opie (eds.), *Oxford Dictionary of Nursery Rhymes* (1951).
82. Charles Calverley. First twenty verses only.
92. H. J. Byron. *Mirth* (1878).
91. Lewis Carroll. Texts vary. This, the fullest, is taken from *The Works of Lewis Carroll*, Roger Lancelyn Green (ed.), (1963).
93, 94. Godfrey Turner. *Mirth* (1878).

95–100. W. S. Gilbert. *The Bab Ballads* (1898).
101. Henry S. Leigh. *Strains from the Strand* (1882).
102, 103. Bret Harte. W. MacDonald (ed.), *Stories and Poems* (1914).
107. E. H. Palmer. Arnold Silcock (ed.), *Verse and Worse* (1952).
110. R. L. Stevenson, *The Moral Emblems* (1921).
111. Anon. 'She Was Poor But She Was Honest.' Auden (ed.), *The Oxford Book of Light Verse* (1938), supplemented by oral collection.
112. A. C. Hilton. From *The Light Green* (n.d.).
113–115. A. D. Godley. **113.** *Lyra Frivola* (1899); **114** and **115.** Fletcher (ed.), *Reliquiae* (1926).
116–121. J. K. Stephen. *Lapsus Calami* (1898).
122, 123. A. E. Housman. **122.** The *Cornhill Magazine*; **123.** *A. E. Housman: Some Poems, Letters and a personal memoir by his brother, Laurence Housman* (1937). **123** is untitled in 'A Selection of Letters' (12 May 1897).
126. Kenneth Grahame. *The Wind in the Willows* (1908).
128. Anon. 'Gasbags' in Silcock (ed.), *Verse and Worse* (1952).
129. Anon. 'The Tale of Lord Lovell' in Carolyn Wells (ed.), *A Parody Anthology* (1904).
130. Anon. 'The Ballad of William Bloat.' Privately communicated.
131. Walter Raleigh. *Laughter from a Cloud* (1923).
132–135. Owen Seaman. **132, 134** and **135.** *Owen Seaman: A Selection* (1937); **130.** verses I–XIII, XXII–XXIV, XXVIII–XXXV; **133.** *Salvage* (1908).
137. B. L. Taylor. From *The Fireside Book of Comic Verse* (1959).
144, 145. Anthony C. Deane, Carolyn Wells (ed.), *A Parody Anthology* (1904).
147. Max Beerbohm. The sub-title has been added.
157. Wallace Irwin. *Letters from a Japanese Schoolboy* (1909).
161. J. C. Squire. *Collected Parodies* (1921), where the following alternative Epilogue is also offered:

> Praise God from whom all blessings flow
> Whose goodness faileth never,
> For men may come and men may go,
> But I go on for ever.

168. Limericks. (i), (iv), (vi), (vii). Baring-Gould (ed.), *The Lure of the Limerick* (1967); (ii), (iii), (v), (viii). *The Oxford Dictionary of Quotations* (1974).
185. Anon. Clerihew, orally communicated.
186. Anon. 'After Emerson' in *Sunday Telegraph* (1975).
187. Stuart Howard-Jones. Privately communicated.
189. Anthony Powell. Privately printed (n.d.).
201. W. H. Auden. I, verses 7–17 and 21–23; III, verses 4–8.
205–207. Wynford Vaughan-Thomas. Privately communicated.
217. Victor Gray. Privately communicated.
234–236. Kingsley Amis. **234.** *Spectator* (1954), under the name Anselm Chilworth.
255. Alan Bennett. Privately communicated.

INDEX OF FIRST LINES

The references are to the numbers of the poems

INDEX OF FIRST LINES

INDEX OF FIRST LINES

INDEX OF FIRST LINES

INDEX OF FIRST LINES

INDEX OF AUTHORS

The references are to the numbers of the poems

INDEX OF AUTHORS